TRANSLATIONAL BIOINFORMATICS AND SYSTEMS BIOLOGY METHODS FOR PERSONALIZED MEDICINE

TRANSLATIONAL BIOINFORMATICS AND SYSTEMS BIOLOGY METHODS FOR PERSONALIZED MEDICINE

QING YAN

Academic Press is an imprint of Elsevier
125 London Wall, London EC2Y 5AS, United Kingdom
525 B Street, Suite 1800, San Diego, CA 92101-4495, United States
50 Hampshire Street, 5th Floor, Cambridge, MA 02139, United States
The Boulevard, Langford Lane, Kidlington, Oxford OX5 1GB, United Kingdom

Library of Congress Cataloging-in-Publication Data
A catalog record for this book is available from the Library of Congress

British Library Cataloguing-in-Publication Data
A catalogue record for this book is available from the British Library

ISBN: 978-0-12-804328-8

For information on all Academic Press publications visit our website at https://www.elsevier.com/books-and-journals

Publisher: Mica Haley
Acquisition Editor: Rafael Teixeira
Editorial Project Manager: Ana Claudia Garcia
Production Project Manager: Lucía Pérez
Designer: Mark Rogers

Typeset by TNQ Books and Journals

CONTENTS

PREFACE

The challenges that healthcare and pharmaceutical industries are facing demand improvements in various aspects, from scientific research to clinical practice. To solve these problems and improve the quality of care, it is urgent to translate the scientific findings from biomedicine into better clinical procedures and results. Because information and knowledge are the major contents in such translational process, novel bioinformatics methodologies such as data integration and knowledge discovery across various domains become critical. As an interdisciplinary field itself, translational bioinformatics provides a special opportunity for overcoming the barriers and obstacles among knowledge domains and clinical branches, and between basic science and clinical bedside practice.

This book provides an introduction and overview of translational bioinformatics and systems biology approaches in support of the development of personalized, systems, and dynamical medicine. The first part of the book introduces and discusses some basic concepts and tools. The second part describes the resources, methods, and applications for finding effective biomarkers and understanding disease complexity. The third part of the book focuses on the translational bioinformatics and systems biology methodologies in drug discovery and clinical applications, including inflammation, cardiovascular diseases (CVDs), cancer, aging, and age-associated diseases.

Specifically, the applications of systems biology and translational bioinformatics may contribute to the development of systems and dynamical medicine with the predictive, preventive, personalized, and participatory (P4) features (see Chapter 2). For the practice of translational bioinformatics, one of the first steps would be to get the necessary resources. Various tools are available for supporting "omics" studies in systems biology (see Chapter 3). Some of the important steps are data integration, data standardization, data mining, knowledge discovery, and decision support (see Chapter 4).

An essential component of personalized medicine is useful biomarkers for quantified and more precise diagnosis and prognosis (see Chapter 5). Proteomics and metabolomics studies are essential in systems biology. The analyses of data from these studies may promote the accuracy, sensitivity, and throughput for biomarker identification because the proteome represents the functional actors in a cell (see Chapter 6). The dynamical properties in the diseases need to be addressed with the shifting targets at various levels during various stages for better therapies (see Chapter 7).

Such approaches would enable the detection and prediction of disease progression and drug responses for improving the safety, utilization, and effects among new and existing drugs, such as the strategies in drug repositioning and drug combinations (see Chapter 8). Translational bioinformatics methods can help identify systems-based biomarkers to address the complexity in the inflammation-associated disease classifiers and patient stratifications (see Chapter 9). Computational systems biology strategies have been proven useful for drug repositioning in the treatment of CVDs (see Chapter 10). The identification of systems-based and dynamical biomarkers representing the evolving processes in cancer development may help support cancer precision medicine (see Chapter 11). Translational bioinformatics may also enhance the understanding in the systems biology of aging with the simulation of the dynamics of biological systems in the aging processes (see Chapter 12).

The integrative and multidisciplinary approaches in the book may be helpful for bridging the gaps among different knowledge domains. This book intends to provide a state-of-the-art and integrative view. By covering topics from basic concepts to novel methodologies, this book can be used by biomedical students, scientific experts, and health professionals at all levels.

Users may include those who are interested in genetics, genomics, proteomics, bioinformatics, systems biology, bioengineering, biochemistry, molecular biology, cell biology, physiology, pathology, microbiology, pharmacology, toxicology, neuroscience, immunology, drug discovery and development, and various branches in clinical medicine.

I would like to thank the editors for their support in this exciting project.

Qing Yan, MD, PhD

PART ONE

Concept and Basic Tools

CHAPTER ONE

Introduction: Translational Bioinformatics and Personalized Medicine

1.1 CURRENT CHALLENGES IN BIOMEDICINE

The tremendous challenges that healthcare and the pharmaceutical industries are facing demand improvements in various aspects, from scientific research to clinical practice. A few examples of these challenges are the rapidly rising costs of clinical care and the growing expenses in drug research and development.

On the other hand, fewer new drugs are being approved by the US Food and Drug Administration, with an increasing rate of high-profile drug withdrawals (Caskey, 2007). In the meantime, the high incidence of adverse drug reactions (ADRs) has become so severe that ADRs are one of the leading causes of morbidity and mortality although many of them are preventable (Ross et al., 2007; Yan, 2011).

Improvements in both scientific and technical aspects are needed to overcome the obstacles and meet the challenges. Considering the scientific aspect, the reductionist drug discovery methods featuring "one-size-fits-all" and single target have been found to contribute to various ADRs (Yan, 2011). These conventional approaches ignore differences between individuals and the interrelationships among drugs, humans, and the environment at various system levels.

In the technological aspect, the gaps in multidisciplinary communications and collaborations have made it difficult to translate the scientific discoveries into more efficient and effective clinical outcomes. In addition, the inadequacies of standardization in the physician ordering systems have led to numerous clinical mistakes and adverse events (Yan, 2010). Another computational challenge related to systems medicine is the integration and analysis of voluminous datasets for identifying patient and disease subtypes (Saqi et al., 2016).

Translational Bioinformatics and Systems Biology Methods for Personalized Medicine
ISBN 978-0-12-804328-8
http://dx.doi.org/10.1016/B978-0-12-804328-8.00001-2

In the scientific aspect, an important factor behind the challenges and obstacles is the conventional healthcare model that is reductionism based and disease centered (Ray, 2004). Such models originating from the late 19th century emphasize the linear bonds between clinical symptoms and pathological detections regarding diseases, diagnosis, and therapeutic approaches (Loscalzo and Barabasi, 2011). On the basis of the reductionist philosophies rather than the complex and nonlinear systems in reality, these simple models are no longer applicable with the novel discoveries in functional genomics and systems biology.

Specifically, approaches such as high-throughput (HTP) technologies and understandings in proteomics, metabolomics, epigenomics, and interactomics have revealed the interrelationships among the components at different system levels (see Chapter 3). Such novel findings request revolutionary improvements in healthcare practice. The novel direction in response to these demands should be heading toward the integrative paradigm that is human centered and individual based (Yan, 2008a).

This change of gear is not possible without scientific and technological support. However, the current situation is that many of the scientific discoveries just stay in the scientific laboratories but cannot benefic clinical practice (Yan, 2010). Although there have been significant scientific advancements, thorough understandings, accurate diagnosis, and effective therapies are still needed for most of the complex diseases.

To solve these problems and improve the quality of care, it is urgent not only to improve but also to translate the scientific findings in biomedicine into better clinical procedures and results (Yan, 2011). The term "translation" here emphasizes the bidirectional flow of information and knowledge between the "bench" side of the basic scientific research and the "bedside" of clinical performance.

Because information and knowledge are the major contents in such translational process, novel bioinformatics methodologies such as data management and knowledge discovery across various domains become critical (see Chapter 4). These approaches would also enable better strategies for drug discovery, development, and administration with lower costs and higher efficiencies.

By addressing the challenges in personalized medicine, translational bioinformatics provides the opportunities and detailed strategies not only for the management and analyses of biomedical data but also for the promotion of proactive and participatory health (Overby and Tarczy-Hornoch, 2013). Translational bioinformatics can serve as the pivotal "vehicle" to integrate various emerging disciplines including pharmacogenomics and systems biology toward the advancement of personalized, preventive, predictive, and participatory (P4) medicine (Hood and Flores, 2012; also see Chapter 2). This chapter will provide an introduction and extensive discussion of this "vehicle."

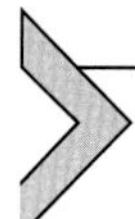

1.2 TRANSLATIONAL BIOINFORMATICS AS THE "VEHICLE" TOWARD PERSONALIZED MEDICINE

1.2.1 The Demand

The advancements in the emerging fields of pharmacogenomics and systems biology may contribute to the development of personalized and systems medicine (Yan, 2008b). As discussed above, this objective is difficult to accomplish without the translational processes bringing the scientific breakthroughs into clinical practices and results. Such translational processes rely on bioinformatics methodologies as the critical "vehicles."

For example, studies in systems biology using technologies such as HTP have generated tremendous amounts of data from both laboratories and clinics. The exponential growth of these data brings both hope and challenges in the storage, management, and analysis to make them ultimately useful for scientific discoveries and disease treatment.

Specifically, conventional information systems do not have the ability to manage and analyze such sizes of data with diversified data types and varied data sources. The high levels of discrepancies make it very difficult to digest and transform them into applicable information and knowledge to enrich both scientific understandings and clinical performances.

Another issue is the ineffective information workflow in the clinical and laboratory settings that have become the obstacles for data sharing and outcome analyses. The conventional information systems can no longer support the current need for data management, data mining, and knowledge discovery (Greenes, 2003).

The information technology itself needs improvements to catch up with the rapidly growing scientific advancements. For instance, at this time most of the experimental and clinical research data are sitting in unconnected servers or stored in different noncompatible databases (Wang et al., 2009). It is very difficult to access or share these data by scientists and clinicians from different groups at different locations. The inefficient communication may block the necessary collaborations across different knowledge domains. The multidisciplinary cooperation is essential for the development of personalized and systems medicine.

In these situations, methods in translational informatics become extremely important to provide the connections between the "bench" studies and the "bedside" practices. The necessary support from translational bioinformatics would allow for the integration of information and knowledge across multiple domains to decode and apply pharmacogenomics and systems biology discoveries into personalized and systems medicine.

It is critical to have comprehensive informatics platforms to support data analysis and visualization for the translational purposes. Via the integration and mining of large patient datasets, systems medicine would enable novel insights into the taxonomy of health and diseases to support personalized intervention schemes (Saqi et al., 2016; also see Chapter 4).

In summary, these challenges, obstacles, and demands have indicated that improvements are needed both scientifically and technically to address the translational steps linking both clinical and laboratory settings. For instance, in the technical aspect, a centralized data warehouse system and the cloud computing technology may be helpful for relieving certain problems. With the integration of scientific and technical improvements, methods in translational bioinformatics may enable more efficient data management and workflow to support better data analysis and decision making in both laboratories and clinics (see Chapter 4).

1.2.2 The Concept

As an independent field, bioinformatics has a history of only a few decades. On the basis of the integrative approaches combining computational methodologies, scientific analysis, and mathematical models, bioinformatics has become indispensable for biomedical studies (Yan, 2003).

Translational bioinformatics may serve as a key subarea of the larger field of "translational medicine" to improve the practice of biomedicine scientifically and clinically including both predictability and outcomes (Day et al., 2009). Translational medicine is an emerging field that combines multiple disciplines and processes to transform biomedical findings into clinical care.

To meet the challenges discussed above, this rapidly advancing discipline is critical to take the role for improving information flow and communications among multiple domains including various scientific areas and clinical branches. In addition to the supporting role, it may also serve as the major player for novel scientific findings and drug discovery by constructing new models and theories.

Specifically, the new methodologies in translational bioinformatics need to work on these tasks scientifically and technically to support the development of personalized and systems medicine:

- more efficient data storage, management, and sharing;
- better data integration and data mining;
- more effective information flow and workflow;
- knowledge discovery and decision support in various settings;
- patient profiling and patient subgroup classification;
- dynamical analyses of both experimental and clinical data; and
- the establishment of systems-based models at various levels with predictive and preventive functions.

As an interdisciplinary field itself, translational bioinformatics provides a special opportunity for overcoming the barriers and obstacles among knowledge domains, clinical branches, scientific findings, and clinical bedside practices. Such efforts rely on the integration of computational biology, bioinformatics, health and medical informatics, genomics and proteomics, systems biology, and various branches in clinical medicine. It may also contribute to the outcome analyses of different intervention methods among different groups of population.

A prominent feature of the development of translational bioinformatics is its evolvement and growth from the simple and rudimentary analysis of molecules such as genes into the complex methodology emphasizing systems biology. As an immediate application, translational bioinformatics has been suggested to improve our understanding of the Human Genome project that may lead to innovative interventions for complex diseases (Sarkar, 2010).

While large-scale biological data can become useful for clinical care, the growing applications of electronic health records (EHRs) may also serve as the valuable suppliers for functional genomics and pharmacogenomics studies. The investigation of genomic data in clinical medicine may bring novel categories of knowledge into the conventional medicine that were previously unavailable. The deep analysis of the copious clinical data would in turn prompt scientific and drug discoveries.

The improved translational processes would enable the applications of genomic technologies for personalized drug repurposing (Denny, 2014; also see Chapter 8). Developments in translational bioinformatics would enable novel algorithms and predictive models for understanding the large-scale datasets and the functional roles of pathways in pathology to support the clinical translation toward personalized medicine (Dauchel and Lecroq, 2016).

1.2.3 The Benefits

An important benefit of translational bioinformatics is the integration of experimental and clinical data streams into more efficient workflow and effective management of resources and expenses. Such improvement would enable better data accessibility, sharing, and exchanging in scientific and clinical settings (Suh et al., 2009). Methods including EHRs and comprehensive knowledge representation would help overcome domain barriers to facilitate collaborations among different groups (Yan, 2010).

The comprehensive approaches in translational bioinformatics may relieve the interoperability issues that have been troubling the biomedical community. Specifically, the enactment and application of data standards would be essential in the translational procedure. The standardization in the computational systems would be crucial for reducing the errors and risks in every step of clinical practice, from physician prescription to drug administration. Such implementations would not only help promote patient satisfaction but also save costs during various phases of health care.

As discussed earlier, a critical obstacle challenging the pharmaceutical industry and healthcare community is the size of scientific and clinical data that is increasing quickly during every step of drug design, discovery, and development (Buchan et al., 2011). With the help from translational bioinformatics, these originally unorganized data can be transformed into precious scientific mines for finding patterns and building predictive models.

The tools of computational biology and health informatics may support the key decision-making procedures from drug development to clinical diagnosis. The novel translational bioinformatics methodologies would allow for the identification of improved drug targets, drug development pipelines, and reduced adverse reactions, with better quality of care.

In summary, these approaches would significantly reduce and prevent adverse events, which is essential for developing safer and personalized medicine. A remarkable benefit of translational bioinformatics is the decision support methodologies that would empower both scientists and clinicians to build personalized profiles and predictive models to bring the right interventions to the right patients (Yan, 2010). These methods would be effective for reducing treatment resistance and adverse reactions. With improved communications, more groups in different disciplines can be involved to make better decisions toward more precise diagnosis and optimized therapeutic outcomes.

1.3 THE GOALS AND MISSIONS

An important task of translational bioinformatics to support personalized medicine is to construct predictive models for disease progression and treatment responses. This task needs to be addressed from the scientific aspect and the informatics (or technological) aspect (Yan, 2010).

Regarding the scientific aspect, translational bioinformatics can work on the improvement of the understanding of complex mechanisms underlying health and diseases. The embracement of the emerging areas, such as complexity theories and systems biology, would empower both biomedical scientists and health professionals to develop more comprehensive diagnostic methods and integrative interventions.

Specifically, various "omics" scientific branches such as transcriptomics and methods such as HTP technologies can be applied to construct patients' biomarker profiles containing various genetic, pathological, and psychological markers in the interrelated networks (Yan, 2011, 2014). Such comprehensive profiles would contribute to the classification of patient subgroups as well as more integrative, reliable and predictive models for diagnosis and prognosis.

Studies in pharmacogenomics and systems biology can become the main support for the scientific aspect of translational bioinformatics with their coverage across various domains, including proteomics, pathology, pharmacology, and clinical medicine (Yan, 2008a,b; also see Chapter 2). Approaches in translational bioinformatics would enable the information flow to overcome the domain barriers with the transformation of unorganized data into actionable knowledge. The integrative applications of HTP technologies, systems biology, and EHRs may lead to a paradigm progress in both clinical care and biomedical sciences (Tenenbaum, 2016).

As a result, the knowledge integration and discovery would improve our understanding of the key issues including the structure–function relations, genes–drugs–environment networks, and genotype–phenotype correlations (Yan, 2008b). A more comprehensive insight into the systemic interrelationships would better explain the dynamical processes and reactions at various system levels. Such more holistic views would be pivotal to more precise diagnosis and therapeutics.

For instance, genetic variations in human kinases have been associated with different illnesses from metabolic disorders to cancers (Lahiry et al., 2010). The understanding of the correlations between structural alterations and the dysfunctions of the kinases as well as pathogenesis may contribute to the findings of more effective drug targets for different diseases. Translational bioinformatics projects such as "The Cancer Genome Atlas" and the "cBioPortal for Cancer Genomics" would promote the translation of cancer biology and genomics into better clinical applications (Sirintrapun et al., 2016; also see Chapter 11).

In the technological and informatics aspect, methods such as workflow integration, data mining, and decision support would be critical for supporting better communications and predictive models (Yan, 2010). These approaches would enable the enactment of standards, which lead to the prevention of errors and adverse events, to promote the quality of health care. The following chapters will provide more detailed discussions from these different aspects in translational bioinformatics for supporting personalized and systems medicine.

REFERENCES

Buchan, N.S., Rajpal, D.K., Webster, Y., Alatorre, C., Gudivada, R.C., Zheng, C., Sanseau, P., et al., 2011. The role of translational bioinformatics in drug discovery. Drug Discov. Today 16 (9–10), 426–434.

Caskey, C.T., 2007. The drug development crisis: efficiency and safety. Annu. Rev. Med. 58, 1–16.

Dauchel, H., Lecroq, T., 2016. Findings from the section on bioinformatics and translational informatics. Yearb. Med. Inf. 207–210.

Day, M., Rutkowski, J.L., Feuerstein, G.Z., 2009. Translational medicine–a paradigm shift in modern drug discovery and development: the role of biomarkers. Adv. Exp. Med. Biol. 655, 1–12.

Denny, J.C., 2014. Surveying recent themes in translational bioinformatics: big data in EHRs, omics for drugs, and personal genomics. Yearb. Med. Inf. 9, 199–205. http://dx.doi.org/10.15265/IY-2014-0015.

Greenes, R.A., 2003. Decision support at the point of care: challenges in knowledge representation, management, and patient-specific access. Adv. Dent. Res. 17, 69–73.

Hood, L., Flores, M., 2012. A personal view on systems medicine and the emergence of proactive P4 medicine: predictive, preventive, personalized and participatory. N. Biotechnol. 29, 613–624.

Lahiry, P., Torkamani, A., Schork, N.J., Hegele, R.A., 2010. Kinase mutations in human disease: interpreting genotype-phenotype relationships. Nat. Rev. Genet. 11 (1), 60–74.

Loscalzo, J., Barabasi, A.-L., 2011. Systems biology and the future of medicine. Wiley Interdiscip. Rev. Syst. Biol. Med. 3 (6), 619–627.

Overby, C.L., Tarczy-Hornoch, P., 2013. Personalized medicine: challenges and opportunities for translational bioinformatics. Per. Med. 10, 453–462.

Ray, O., 2004. The revolutionary health science of psychoendoneuroimmunology: a new paradigm for understanding health and treating illness. Ann. N.Y. Acad. Sci. 1032, 35–51.

Ross, C.J.D., Carleton, B., Warn, D.G., Stenton, S.B., Rassekh, S.R., Hayden, M.R., 2007. Genotypic approaches to therapy in children: a national active surveillance network (GATC) to study the pharmacogenomics of severe adverse drug reactions in children. Ann. N.Y. Acad. Sci. 1110, 177–192.
Saqi, M., Pellet, J., Roznovat, I., Mazein, A., Ballereau, S., De Meulder, B., Auffray, C., 2016. Systems medicine: the future of medical genomics, healthcare, and wellness. Methods Mol. Biol. (Clifton N.J.) 1386, 43–60. http://dx.doi.org/10.1007/978-1-4939-3283-2_3.
Sarkar, I.N., 2010. Biomedical informatics and translational medicine. J. Transl. Med. 8, 22.
Sirintrapun, S.J., Zehir, A., Syed, A., Gao, J., Schultz, N., Cheng, D.T., 2016. Translational bioinformatics and clinical research (biomedical) informatics. Clin. Lab. Med. 36 (1), 153–181. http://dx.doi.org/10.1016/j.cll.2015.09.013.
Suh, K.S., Remache, Y.K., Patel, J.S., Chen, S.H., Haystrand, R., Ford, P., Shaikh, A.M., et al., 2009. Informatics-guided procurement of patient samples for biomarker discovery projects in cancer research. Cell Tissue Bank. 10 (1), 43–48.
Tenenbaum, J.D., 2016. Translational bioinformatics: past, present, and future. Genom. Proteom. Bioinform. 14 (1), 31–41. http://dx.doi.org/10.1016/j.gpb.2016.01.003.
Wang, X., Liu, L., Fackenthal, J., Cummings, S., Olopade, O.I., Hope, K., Silverstein, J.C., et al., 2009. Translational integrity and continuity: personalized biomedical data integration. J. Biomed. Inform. 42 (1), 100–112.
Yan, Q., 2003. Bioinformatics and data integration in membrane transporter studies. Methods Mol. Biol. (Clifton N.J.) 227, 37–60.
Yan, Q., 2008a. Pharmacogenomics in drug discovery and development. Preface. Methods Mol. Biol. (Clifton N.J.) 448, v–vii.
Yan, Q., 2008b. The integration of personalized and systems medicine: bioinformatics support for pharmacogenomics and drug discovery. Methods Mol. Biol. (Clifton N.J.) 448, 1–19.
Yan, Q., 2010. Translational bioinformatics and systems biology approaches for personalized medicine. Methods Mol. Biol. (Clifton N.J.) 662, 167–178.
Yan, Q., 2011. Toward the integration of personalized and systems medicine: challenges, opportunities and approaches. Per. Med. 8, 1–4.
Yan, Q., 2014. From pharmacogenomics and systems biology to personalized care: a framework of systems and dynamical medicine. Methods. Mol. Biol. (Clifton, NJ) 1175, 3–17.

CHAPTER TWO

Systems and Dynamical Medicine: The Roles of Translational Bioinformatics

2.1 THE INTEGRATION OF PHARMACOGENOMICS AND SYSTEMS BIOLOGY

The emerging fields of pharmacogenomics and systems biology may enable fundamental advancements and serve as the scientific basis that can be translated into personalized and systems medicine (Yan, 2008a,b, 2014). The rapidly developing discipline of pharmacogenomics may enrich our knowledge in genomics for understanding the individual variances in response to medications and vaccines (Yan, 2008a,b).

With the understanding of the genetic discrepancies and genomic pools, pharmacogenomics may provide promising discoveries for the prediction of disease predispositions and treatment outcomes (Meyer, 2004). Such findings would be critical for the prevention of disease progression and adverse reactions for the optimal therapies and reduced costs.

However, the focus of pharmacogenomics is not just to identify single nucleotide polymorphisms or isolated disease markers. Genes are always interacting with other components in the complex networks rather than carrying out solitary performances. The interacting components in the complex network include other molecules and chemicals such as drugs, vaccines, and various environmental elements. Therefore, the integration of pharmacogenomics and systems biology is essential for the relevant scientific discoveries to be translated into systems and personalized medicine (Yan, 2011b).

Focusing on the interrelationships at various system levels and scales, systems biology provides a comprehensive view of health and diseases. The multiple dimensions cover both spatial and temporal aspects, including those from molecular to cellular and organ levels, as well as the scales from seconds and minutes to days and months (Yan, 2010; also see Chapter 7). Such perceptive would enable the integration of information about genotypes,

Translational Bioinformatics and Systems Biology Methods for Personalized Medicine
ISBN 978-0-12-804328-8
http://dx.doi.org/10.1016/B978-0-12-804328-8.00002-4

phenotypes, environment, and dynamics to improve conceptual studies and clinical practice (Sarkar et al., 2011).

By linking structural concepts with functional and dynamical behaviors of biomedical systems, studies in systems biology are undertaking the mission to understand the interactions and networks as a whole, rather than isolated components. The holistic approaches would contribute to the thorough understanding of the malfunctions of the human system and the "root" of diseases on the individual basis (Yan, 2011b). In summary, the combination of both pharmacogenomics and systems biology is needed to pave the scientific ground toward the clinical practice of personalized, systems, and dynamical medicine (Yan, 2014).

2.2 TRANSLATIONAL BIOINFORMATICS, PERSONALIZED AND SYSTEMS MEDICINE

Systems biology and systems medicine may address the complete pathological processes from disease onsets to progressions with the highlight of the overall pathway kinetics (Ayers and Day, 2015). Such strategies may help elucidate the dynamical interactive networks in the course of medical conditions for finding clinical targets toward more precise diagnosis and therapeutics. With the investigation of fundamental network biology, targeted proteomics may be linked to the phenotypic conditions to support more advanced biomarker discovery and validation for complex diseases such as cardiovascular disease and cancer (Ebhardt et al., 2015).

To achieve these goals, the scientific efforts would need strong support from advanced technologies. Challenges still need to be met for the efficient analysis of data from high-throughput technologies, chromatography, mass spectrometry, and nuclear magnetic resonance. The application of translational bioinformatics would promote the advancement of systems biology by incorporating the data from studies of genomes, proteomes, transcriptomes, metabolomes, and epigenomes (Auffray et al., 2009; also see Chapter 3).

Translational bioinformatics approaches such as data integration and mining can be applied to understand the complex datasets from both population groups and individuals (see Chapter 4). Such approaches would enable the understanding of both structural and functional interrelationships for better biomarkers and therapeutic target selections (see Chapters 5 and 6). These methods would be critical for systems biology to overcome the data obstacles and make major advancements in the comprehensive understanding of health and illnesses.

With the integration of pharmacogenomics and systems biology, an important advancement may be achieved as the potential revolution in biomedicine, that is, the transformation from the conventional reductionism-grounded and disease-centered biomedical framework to a dynamical systems-established and human-centered model (Yan, 2011a).

Many properties of pharmacogenomics and systems biology may serve as the scientific root for the development of personalized and systems medicine. One of such properties is the multidisciplinary connections that would enable a large-scale interpretation of health and diseases across different knowledge domains (Yan, 2008b). These multiple domains contain the biological aspects, including proteomics and physiology, as well as clinical medical aspects, such as epidemiology and internal medicine.

Translational bioinformatics also has a pivotal role enabling quantitative assessments, mathematical analyses, and predictive models. These features would allow for across-the-board models monitoring the information flow in various knowledge domains. Such models would be helpful for understanding the dynamics in the complex adaptive systems (CASs) in various states of health and illnesses. As shown in Fig. 2.1, the essential features of

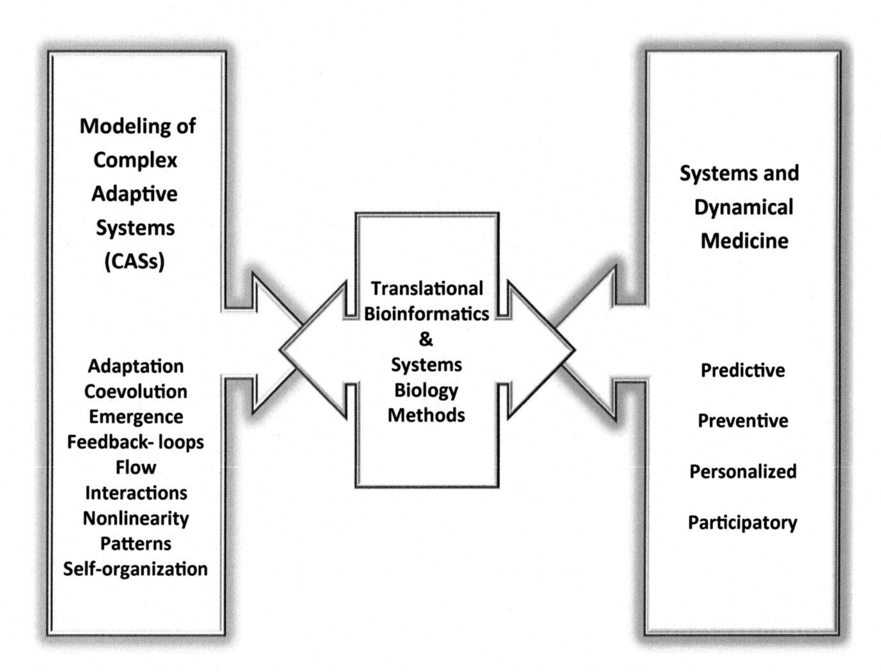

Figure 2.1 The modeling of complex adaptive systems (CASs) for systems and dynamical medicine.

CASs contain the concepts of emergence, adaptation, self-organization, and nonlinearity (Chaffee and McNeill, 2007).

The understanding of these features of CASs would be critical for making the revolutionary advancements toward personalized and systems medicine. The elucidation of the complex interconnections and dynamical networks in the human system is the key for constructing predictive and preventive models for better diagnosis and interventions.

2.3 THE BASIC CONCEPTS OF THE COMPLEX "WHOLE BODY SYSTEM"

2.3.1 Emergence and Interaction Patterns: Human-Centered Medicine

As one of the most basic concepts of the CASs, emergence is a key characteristic of complex systems. An important perception of a holistic system is that "the whole is greater than the sum of its parts" (Yan, 2010). In a complex biomedical system, the actions or signs are the collective consequences or effects evolving or "emerging" from the interconnections and dynamical interactions across various scales (Chaffee and McNeill, 2007; Iris, 2008). The elements or entities are referred to as the acting "agents" of the complex systems. An important task of systems biology is to elucidate these interrelationships, interactions, and the dynamical processes in the coevolution of the agents.

Conventional reductionist models cannot explain or predict such features on the basis of single components or separated and static segments. For instance, the overall cellular phenotypes are the emergent results arising from the overall nonlinear interrelationships among a network of microenvironmental elements (Dinicola et al., 2011), such as the communications between the mitochondria and nucleus. Another example is the occurrence of systemic inflammation as the consequence of various feedbacks and pathways including the microbiota–gut–brain axis, rather than the product of isolated cytokines (see Chapter 9).

Systems biology studies would help identify the patterns of interrelationships for the better understanding of the features of emergence in health and diseases. Specifically, it is often observed that different arrangements of agents such as drugs may reach the same consequences or effects, but the similar grouping of agents may still have different results in different patients (Sturmberg and Martin, 2013). Such phenomena emphasize the decisive roles of the functional interactions instead of the isolated structural components for individualized therapies.

The understanding of the concept of emergence and the underlying mechanisms is critical for making revolutionary progress in systems and personalized medicine that is quite different from the conventional reductionist paths. As an example, although cancer has been considered as one type of the disease, the underlying mechanisms have referred to different arrays of genetic variations and pathways in different tissues and organs among various patients (Bleeker et al., 2009). Such discrepancies request for personalized treatments at different stages.

In another example, although different patients may have different illnesses, such as obesity, type 2 diabetes, and heart diseases, they may have the common cause such as the similar lifestyle and unhealthy dietary habits. Studies have confirmed that chronic inflammation may be the shared mechanism among the different chronic disorders such as diabetes, cardiovascular diseases, kidney problems, Alzheimer's disease, and cancer (Manabe, 2011; also see Chapter 9). Recent discoveries have revealed that many apparently unrelated illnesses may be responsive to the same treatment such as IL 1β neutralization, including gout, type 2 diabetes, heart failure, and cancer (Dinarello, 2011).

Such elucidations may change the conventional strategies of "one drug for one disease". Instead, personalized and systems medicine would consider the "whole human body system" as the illness "root". With this change of concept, the same type of disease needs different therapies for different patients based on the varied etiology and stages. However, if different illnesses have the similar mechanisms, e.g., shared inflammatory pathways, the similar interventions may be applied to different patients.

Such strategies would enable the transformation or evolution from the disease-centered medicine to human-centered health care. In addition, the similar treatments can be designed for different illnesses with the shared mechanisms by expanding the administration from the currently available drugs. Such drug repositioning approaches would be cost effective because they would enable more efficient applications of the existed drugs, saving the expenses for both basic research and clinical treatments (see Chapter 8).

2.3.2 Adaptation and Coevolution: The Dynamical Processes

In addition to emergence, "adaptation" is also a basic concept of CASs (Chaffee and McNeill, 2007). A remarkable feature of CASs is that they have the ability of adaption to environmental changes and stresses. Under novel conditions, the agents of the CASs may adapt and evolve. Same as health, disease is also always in dynamical conditions. The dynamical activities of diseases are manifested as the progressive signs during different stages of the disease development. Such changes represent the evolvement that the "whole body system" is in the process of adapting to the alterations in the environment. The signs of such "adaptation" at various disease stages reflect the ongoing dynamical communications in the organism that cannot be assessed with a single clinical factor or separated causes (Heng, 2008).

For example, the processes of adaptation can be detected in many illnesses such as coronary artery disease, chronic obstructive airway problems, and rheumatoid arthritis (Sturmberg and Martin, 2013). The conditions of dynamical coevolution may be a feature of the relevant activities such as those shown in disease progression. Because each of the interacting "agents" such as the signaling pathways may be changing or evolving in such adaptive processes, different dynamics and signs may emerge in subsystems.

Specifically, drug resistance (e.g., against multiple drugs) is one of the most significant problems in treating various illnesses, especially infectious diseases and cancers (Avner et al., 2012). Such problems can be addressed with the understanding of the "adaptive" activities in CASs. That is, the pathogenic adaptation may be the underlying mechanism accounting for such obstacles.

The concept of adaptation is very useful for overcoming the drug resistance obstacles by applying different treatment strategies at different time points or stages for even the same patient or patients with the same diseases. This is also a key point for the practice of personalized medicine. As a common feature of CASs, the processes of adaptation can be targeted in both chronic diseases, including cancer, and acute diseases, such as viral infections, to improve the treatment effects.

2.3.3 Self-Organization and Feedback Loops: The Robust Networks

To have a better understanding of how the CASs work in health and diseases, the concepts of self-organization and robustness are also essential. The functions of CASs rely on multidirectional interactions and positive and negative feedback loops at different levels or scales. However, these mechanisms can function themselves and do not need outside guidance or higher level instructions (Chaffee and McNeill, 2007; Iris, 2008).

To illustrate such features, some well-known examples in biological organisms are the maintenance of the homeostasis of water, body temperature, blood pressure, and blood glucose. In the healthy state, various levels of robustness can be observed in response to environmental changes such as seasonal alterations. Even in the disease conditions, the pathogenic mechanisms may develop certain robustness against drug therapies (Kitano, 2007).

The elucidation of these features of the living CASs is especially meaningful for the development of personalized medicine. That is, the purpose of treatments can now focus on the correction and "tuning" of the abnormal status back to the normal healthy condition, rather than just "winning" the battle of diseases by killing the pathogens or relevant cells.

Use infectious diseases as an example. The features of CASs exist in both hosts and pathogens, and the host–pathogen interactions should be the essential treatment targets, but not the bacteria or viruses alone as in the conventional medicine. These features decide that simple methods for eliminating the bacteria or viruses using antibiotics would result in drug resistance and various side effects.

To solve the problems, interventions targeting the dynamical host–pathogen interactions are more appropriate. Such interventions should be comprehensive to tackle the positive and negative feedback loops and the dynamical pathways. Various factors and conditions should be considered, including the relative constancy resulted from self-organization and robustness, as well as the elasticity from adaptation.

On the basis of such understanding, an emphasis of the treatment targets can be put on multiple inflammatory pathways during different stages involving both host and pathogenic molecules, but no longer just simply the pathogens. Such a "switch of the gear" allows for the transformation of disease-centered medicine to human-centered care.

More importantly, the feedback loops should have the crucial roles in the potential targets as they provide the dynamical connections among the interactive elements such as proteins, drugs, behaviors, and environment at different system levels. The understanding of such structure–function and genotype–phenotype associations at various levels would be pivotal for more precise diagnosis and treatment (Yan, 2008b).

Specifically, structural and functional variations at the molecular level and protein–drug interactions may have further influences on the networks and pathways at the cellular and system levels (Yan, 2012). Meanwhile, the communal consequences from the "emergent" features arising from these interactions can be shown as clinical signs and manifestations, disease activities, as well as treatment responses at the organismal level. The comprehension of such complex interrelationships at various system levels may provide more integrative methods for the early prediction, detection, diagnosis, and prevention of diseases.

Furthermore, the interactions at each level may not just be apparent signs but also have real biophysical and biochemical impacts on other levels. The higher level activities may affect the lower level processes and vice versa. For instance, stress including sunburn at the environmental level may result in DNA damages among skin cells with alterations at both molecular and cellular levels.

Therefore, a more complete model would be needed to detect the interactions not just within the same level, such as the protein–protein communications, but also between and across various levels, such as the genes–cells–environment correlations (Qu et al., 2011). A more systemic view would be possible by embracing the across-level feedback loops and pathways including the genotype–phenotype correlations at different temporal and spatial scales (see Chapter 7). Systems biology models based on such holistic views would revolutionize both biological research and medical practice from isolated symptoms to systems-based biomarkers, from the failed "one-drug-fits-all" method to personalized and more effective care (Yan, 2008a,b).

2.3.4 Nonlinearity and Dynamical Pathophysiology

In addition to the concepts discussed previously, a prominent feature of CASs is that proportional reactions or consequences may not be achieved from the original stimulus. Meanwhile, the system may show high sensitivities to the initial condition with the potential occurrence of enormous alterations (Chaffee and McNeill, 2007). Such phenomena are called "nonlinearity".

The feature of nonlinearity can be observed in many clinical conditions, demonstrating the importance of understanding such concepts in the human CASs. A well-known example is that in chemotherapy of cancers, therapeutic results cannot be improved just by increasing the dosages because the dosage–outcome relationships are not linear (Leyvraz et al., 2008). It is often seen that initial chemotherapies can be effective in shrinking the sizes of tumors. However, continuous and higher dosages may also cause the generation of secondary tumors (Mittra, 2007). Another example is in cardiac electrophysiology. Nonlinearities have been associated with the cardiac arrhythmogenesis with critical roles in maintaining cardiac rhythms (Krogh-Madsen and Christini, 2012).

For the achievement of personalized medicine, it is essential to develop prevention and treatment strategies based on individualized medications, dosages, intensities, timing, and frequencies at different phases of the diseases, that is, "to bring the right interventions to the right people with the right dosages and intensities at the right time" (Yan, 2014). These goals cannot be accomplished without incorporating the features of CASs especially the effects of nonlinearity.

CASs are open systems interacting constantly with the environment. This is also true for the human body that communicates with both natural and social surroundings continually. Substantial alterations may occur quickly from the nonlinear human–environment connections (Sturmberg and Martin, 2013). Various CASs factors may be involved in such processes in addition to nonlinearity, including timing, dynamics, adaptation, as well as feedback loops.

For instance, seasonal changes have been closely associated with various infectious diseases including influenza. Different host–pathogen–environment interactions and pathways are involved. Social incidents including attacks and battles may lead to serious psychological complications including posttraumatic stress disorder. Preventive strategies in personalized and systems medicine need to incorporate these nonlinear biopsychosocial and environmental factors.

The previous examples also indicate that a remarkable feature of CASs is the dynamical variations in the flow of substances, energy, and information across various temporal scales and spatial levels (see Chapter 7). It is pivotal to discern these features because health and illnesses have dynamical and adaptive routes rather than stagnant or inactive conditions. The understanding of normal physiology relies on such properties of nonlinear dynamics. Similarly, the perceptions of the pathophysiology of diseases also depend on the detections of the dynamical changes (Buchman, 2004).

As an example in cardiology, the sinus rhythm needs to be detected for the regularly recurring dynamics, and atrial fibrillation can be assessed for the irregularly recurring dynamics (Chay and Rinzel, 1985). Effective diagnostic and treatment strategies need to detect the constantly changing physiological and pathological factors in the same patient at different time points, as well as among different patients during certain periods. Based on the comprehensive models incorporating these factors, systems and dynamical medicine can be developed for the optimal outcomes.

2.4 SYSTEMS AND DYNAMICAL MEDICINE WITH P4 FEATURES

As illustrated in Fig. 2.1, the applications of systems biology and translational bioinformatics may contribute to the development of systems and dynamical medicine with the predictive, preventive, personalized, and participatory (P4) features (Younesi and Hofmann-Apitius, 2013; Hood and Flores, 2012). Systems-based and dynamical models focusing on the features of CASs may provide the revolutionary resolutions to overcome both theoretical and practical obstacles in conventional medicine.

For instance, the conventional therapeutic guidelines for colorectal cancer put emphasis on clinical characteristics including cancer stages and grades. However, comprehensive and robust biomarkers would be more useful by addressing driver mutations, signaling proteins, microRNAs, as well as long noncoding RNAs (Castagnino et al., 2016). Integrative models and profiles embracing the signaling networks with dynamical predictions may help identify drug combination priorities and reduce the number of drugs to be examined for better clinical outcomes (see Chapter 8).

The systems-based dynamical approaches would empower both scientists and clinicians to find better therapeutic targets for various types of complex diseases at different stages. An important step in such efforts would be the discovery of systems-based and dynamical biomarkers for the timely alerts for presymptomatic diagnosis and prognosis to support the prediction and prevention of diseases during various phases (Bengoechea, 2012; also see Chapter 5). Such biomarkers would be helpful for decreasing the risks for disease occurrence and disability, which is the key for preventive medicine in at-risk populations (Younesi and Hofmann-Apitius, 2013). These strategies may be particularly useful for chronic and complex illnesses, including cancer and Alzheimer's disease. More detailed discussions on systems-based and dynamical biomarkers will be available in Chapters 5–12.

Moreover, the detection of the alterations across various spatial levels and temporal scales may allow for the finding of the evolving treatment targets in personalized medicine (see Chapter 7). More precise and robust biomarkers can be applied for disease stratification and patient subgroups identification for more individualized interventions.

Furthermore, the elucidation of the complex human–drug interactions would enable the prediction of therapeutic responses to reduce adverse reactions and to improve clinical results. The systems-based and dynamical disease predictive models can be built by analyzing various "omics" data for the transformation from after-disease reactive interventions to proactive care. Such models would allow for the changes from disease-centered to human-centered care to promote the involvement of individuals to achieve the objective of participatory medicine.

REFERENCES

Auffray, C., Chen, Z., Hood, L., 2009. Systems medicine: the future of medical genomics and healthcare. Genome Med. 1 (1), 2.

Avner, B.S., Fialho, A.M., Chakrabarty, A.M., 2012. Overcoming drug resistance in multi-drug resistant cancers and microorganisms: a conceptual framework. Bioengineered 3, 262–270.

Ayers, D., Day, P.J., 2015. Systems medicine: the application of systems biology approaches for modern medical research and drug development. Mol. Biol. Int. 2015, 698169.

Bengoechea, J.A., 2012. Infection systems biology: from reactive to proactive (P4) medicine. Int. Microbiol. 15, 55–60.

Bleeker, F.E., Lamba, S., Rodolfo, M., et al., 2009. Mutational profiling of cancer candidate genes in glioblastoma, melanoma and pancreatic carcinoma reveals a snapshot of their genomic landscapes. Hum. Mutat. 30, E451–E459.

Buchman, T.G., 2004. Nonlinear dynamics, complex systems, and the pathobiology of critical illness. Curr. Opin. Crit. Care 10, 378–382.

Castagnino, N., Maffei, M., Tortolina, L., et al., 2016. Systems medicine in colorectal cancer: from a mathematical model toward a new type of clinical trial. Wiley Interdiscip. Rev. Syst. Biol. Med. 8, 314–336.

Chaffee, M.W., McNeill, M.M., 2007. A model of nursing as a complex adaptive system. Nurs. Outlook 55, 232–241.

Chay, T.R., Rinzel, J., 1985. Bursting, beating, and chaos in an excitable membrane model. Biophys. J. 47, 357–366.

Dinarello, C.A., 2011. Blocking interleukin-1β in acute and chronic autoinflammatory diseases. J. Intern. Med. 269, 16–28.

Dinicola, S., D'Anselmi, F., Pasqualato, A., et al., 2011. A systems biology approach to cancer: fractals, attractors, and nonlinear dynamics. OMICS 15, 93–104.

Ebhardt, H.A., Root, A., Sander, C., Aebersold, R., 2015. Applications of targeted proteomics in systems biology and translational medicine. Proteomics 15, 3193–3208.

Heng, H.H.Q., 2008. The conflict between complex systems and reductionism. JAMA 300, 1580–1581.

Hood, L., Flores, M., 2012. A personal view on systems medicine and the emergence of proactive P4 medicine: predictive, preventive, personalized and participatory. N. Biotechnol. 29, 613–624.

Iris, F., 2008. Biological modeling in the discovery and validation of cognitive dysfunctions biomarkers. In: Turck, C.W. (Ed.), Biomarkers for Psychiatric Disorders. Springers Science + Business Media, New York.

Kitano, H., 2007. The theory of biological robustness and its implication in cancer. Ernst Scher. Res. Found. Workshop 69–88.

Krogh-Madsen, T., Christini, D.J., 2012. Nonlinear dynamics in cardiology. Annu. Rev. Biomed. Eng. 14, 179–203.

Leyvraz, S., Pampallona, S., Martinelli, G., et al., 2008. A threefold dose intensity treatment with ifosfamide, carboplatin, and etoposide for patients with small cell lung cancer: a randomized trial. J. Natl. Cancer Inst. 100, 533–541.

Manabe, I., 2011. Chronic inflammation links cardiovascular, metabolic and renal diseases. Circ. J. 75, 2739–2748.

Meyer, U.A., 2004. Pharmacogenetics – five decades of therapeutic lessons from genetic diversity. Nat. Rev. Genet. 5 (9), 669–676.

Mittra, I., 2007. The disconnection between tumor response and survival. Nat. Clin. Pract. Oncol. 4, 203.

Qu, Z., Garfinkel, A., Weiss, J.N., Nivala, M., 2011. Multi-scale modeling in biology: how to bridge the gaps between scales? Prog. Biophys. Mol. Biol. 107, 21–31.

Sarkar, I.N., Butte, A.J., Lussier, Y.A., Tarczy-Hornoch, P., Ohno-Machado, L., 2011. Translational bioinformatics: linking knowledge across biological and clinical realms. J. Am. Med. Inf. Assoc. 18 (4), 354–357.

Sturmberg, J.P., Martin, C.M., 2013. Complexity in health: an introduction. In: Sturmberg, J.P., Martin, C.M. (Eds.), Handbook of Systems and Complexity in Health. Springer Science + Business Media, New York.

Yan, Q., 2008a. Pharmacogenomics in drug discovery and development. Preface. Methods Mol. Biol. (Clifton, NJ) 448, v–vii.

Yan, Q., 2008b. The integration of personalized and systems medicine: bioinformatics support for pharmacogenomics and drug discovery. Methods Mol. Biol. 448, 1–19.

Yan, Q., 2010. Translational bioinformatics and systems biology approaches for personalized medicine. Methods Mol. Biol. (Clifton, NJ) 662, 167–178.

Yan, Q., 2011a. Translation of psychoneuroimmunology into personalized medicine: a systems biology perspective. Pers. Med. 8, 641–649.

Yan, Q., 2011b. Toward the integration of personalized and systems medicine: challenges, opportunities and approaches. Pers. Med. 8, 1–4.

Yan, Q., 2012. The role of psychoneuroimmunology in personalized and systems medicine. Methods Mol. Biol. 934, 3–19.

Yan, Q., 2014. From pharmacogenomics and systems biology to personalized care: a framework of systems and dynamical medicine. Methods Mol. Biol. (Clifton, NJ) 1175, 3–17.

Younesi, E., Hofmann-Apitius, M., 2013. From integrative disease modeling to predictive, preventive, personalized and participatory (P4) medicine. EPMA J. 4, 23.

CHAPTER THREE

Translational Bioinformatics Support for "Omics" Studies: Methods and Resources

3.1 INTRODUCTION

As discussed in Chapters 1 and 2, the tremendous amount of data from both experimental and clinical studies and the complexity in biomedical systems are referring to the key role of translational bioinformatics. Translational bioinformatics is indispensable for the integration of various knowledge domains in pharmacogenomics and systems biology for the translation of basic research into personalized, systems, and dynamical medicine (Yan, 2010; also see Chapters 1 and 2).

Translational bioinformatics may help to overcome the barriers between the "omics" domains in systems biology and provide the interpretation for the results from high-throughput (HTP) analyses. These applications are critical for the establishment of patients' profiles and subgroups in personalized medicine (Yan, 2012).

Specifically, approaches such as data integration and data mining are not only useful for knowledge discovery (KD) (see Chapter 4) but can also contribute to the construction of predictive models by addressing the complex adaptive features of the biomedical systems and dynamical pathways for more effective disease prevention and treatment (see Chapter 2). These methods are essential for the elucidation of the complex activities and processes in health and diseases across various spatial levels and temporal scales (see Chapter 7).

In addition to preventive and therapeutic strategies, translational bioinformatics may also enable the identification of more precise and robust biomarkers to improve diagnosis and prognosis (see Chapter 5). Applications in translational bioinformatics may improve the efficacy in almost every step of biomedical research and clinical practice to empower both scientists and clinicians for the practice of human-centered systems and dynamical medicine (see Chapters 1 and 2).

Translational Bioinformatics and Systems Biology Methods for Personalized Medicine
ISBN 978-0-12-804328-8
http://dx.doi.org/10.1016/B978-0-12-804328-8.00003-6

Specifically, the integration of various "omics" analyses (e.g., proteomics and transcriptomics) with clinical data assessments in both biological and healthcare informatics may especially be helpful for understanding across-level associations. Such associations are pivotal for the establishment of systems medicine, especially the gene–drug–environment interactions and genotype–phenotype relations (Yan, 2010; also see Chapter 2). Detailed resources, tools, and approaches of translational bioinformatics for these purposes will be discussed in the subsequent sections in this chapter.

In addition, data integration approaches may connect laboratory and clinical data streams to support better workflow and collaborations in both research and clinical settings (Suh et al., 2009). Methods including data mining and electronic health records (EHRs) may promote more powerful decision support for drug design and therapeutic strategies to enable the right prevention and treatment to the right patients at the right time (Yan, 2010; also see Chapter 2). Translational bioinformatics needs to combine both bioinformatics and health informatics methods to lower the healthcare costs, risks, and adverse reactions. These integrative and decision support methods will be discussed in more detail in Chapter 4.

3.2 BIOINFORMATICS METHODS AND RESOURCES FOR "OMICS" STUDIES

For the practice of translational bioinformatics, one of the first steps would be to get the necessary resources. Many resources and tools can be applied for the development of systems and dynamical medicine (see Chapters 1 and 2). Some of the databases and tools are listed in Tables 3.1–3.4. With more and more works in the areas, such kind of lists are always growing.

As illustrated in Fig. 3.1, various tools are available for supporting "omics" studies in systems biology (Halberg et al., 2007), including:

- genomics;
- proteomics;
- epigenomics;
- transcriptomics;
- metabolomics;
- lipidomics;
- pharmacogenomics; and
- chronomics.

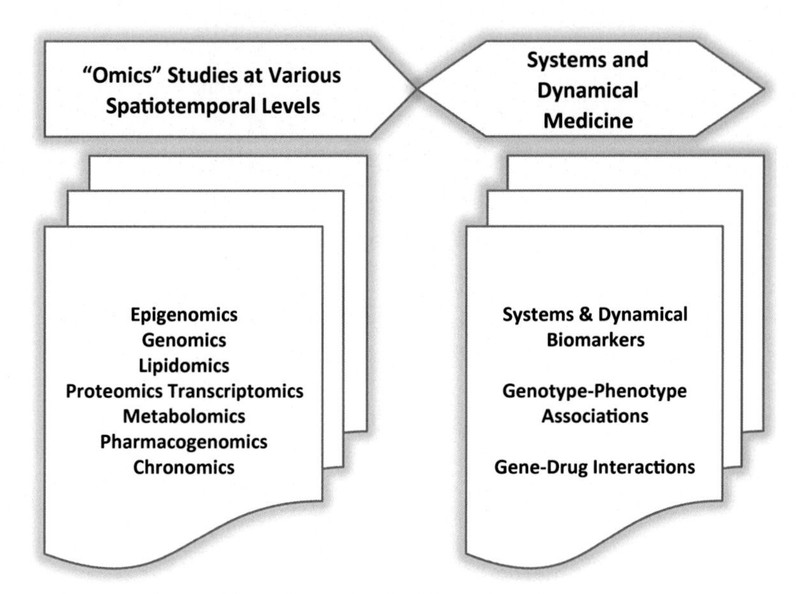

Figure 3.1 Translational bioinformatics for "Omics" studies to support systems and dynamical medicine.

Specifically, genomics and proteomics are essential for understanding the structure–function associations at the molecular level. Various sources provide such information; many of them are comprehensive platforms with cross-database search engines (see the URLs in Table 3.1). These resources include:

- National Center for Biotechnology Information (NCBI): provides access to databases about DNA, RNA, proteins, domains, structures, expression, maps, variations, etc.;
- Ensembl: a genome browser for comparative genomics, expression, sequence variations, multiple alignments, regulatory functions, diseases, etc.;
- ExPASy: a resource portal that provide information about genomics, proteomics, structure analysis, systems biology, evolutionary biology, population genetics, transcriptomics, glycomics, medicinal chemistry, etc.;
- Uniprot: a central repository of protein sequences and functions.

Based on the genetics databases, analyses can be performed for sequence similarities, structures, motifs, patterns, phylogenetic trees, functions, interactions, and evolutionary changes. As listed in Table 3.1, these tools include:

Table 3.1 Bioinformatics Resources and Tools for "Omics" Studies

Tools	Web URLs	Contents
ArrayExpress	http://www.ebi.ac.uk/arrayexpress/	Gene expressions
Biomarkers and Systems Medicine (BSM)	http://pharmtao.com/health/category/systems-medicine/biomarkers-systems-medicine/	Biomarkers
BLAST	http://blast.ncbi.nlm.nih.gov/Blast.cgi	Sequence similarities
Cancer Proteomics Database	http://cancerproteomics.uio.no/	Cancer proteomics
CLUSTAL	http://www.clustal.org/	Sequence alignments
Cytoscape	http://www.cytoscape.org	Complex networks
dbSNP	http://www.ncbi.nlm.nih.gov/projects/SNP/	Single nucleotide polymorphisms
Ensembl	http://uswest.ensembl.org/index.html	Genomics
ExPASy	http://expasy.org/	Bioinformatics
Gene Expression Omnibus (GEO)	http://www.ncbi.nlm.nih.gov/geo/	Gene expressions
The Human Metabolome Database (HMDB)	http://www.hmdb.ca/	Small molecule metabolites
Human Protein Atlas (HPA)	http://www.proteinatlas.org/	Protein expressions
The Human Protein Reference Database (HPRD)	http://www.hprd.org/	Pathways and proteins
IntAct	http://www.ebi.ac.uk/intact/	Molecular interactions
isoMETLIN	https://isometlin.scripps.edu/	Metabolomics
Kyoto Encyclopedia of Genes and Genomes (KEGG)	http://www.genome.jp/kegg/pathway.html	Pathways
LipidHome	http://www.ebi.ac.uk/apweiler-srv/lipidhome	Lipidomics
LIPID MAPS Lipidomics Gateway	http://www.lipidmaps.org/	Lipidomics
MIPS	http://mips.helmholtz-muenchen.de/proj/ppi/	Protein–protein interactions
MitoMiner	http://mitominer.mrc-mbu.cam.ac.uk/release-4.0/begin.do	Mitochondrial proteomics

Table 3.1 Bioinformatics Resources and Tools for "Omics" Studies—cont'd

Tools	Web URLs	Contents
National Center for Biotechnology Information (NCBI)	http://www.ncbi.nlm.nih.gov/	A cross-database search engine
The Protein Data Bank (PDB)	http://www.wwpdb.org/	Protein structures
PROSITE	http://prosite.expasy.org/	Protein families
Reactome	http://www.reactome.org/	Pathways
RNA-Seq Atlas	http://medicalgenomics.org/rna_seq_atlas	Gene expressions
Small Molecule Pathway Database (SMPDB)	http://smpdb.ca/	Human small molecule pathways
The 1000 Genomes Project	http://www.1000genomes.org/	Sequence variations
The Human Variome Project	http://www.humanvari-omeproject.org/	Genetic variations and health
UniProt	http://www.uniprot.org/	Proteins

- BLAST (Altschul et al., 1990): similarities among nucleotide or protein sequences;
- PROSITE (Sigrist et al., 2010): protein domains, families, functional sites, patterns, and profiles;
- CLUSTAL (Larkin et al., 2007): multiple alignments of nucleic acid and protein sequences; and
- The Protein Data Bank (PDB) (Rose et al., 2011): structures of proteins and nucleic acids for searching and visualizing.

The assessment of genetic variances such as single nucleotide polymorphisms (SNPs) is particularly useful for pharmacogenomics studies to examine and predict individual variations in drug responses for personalized medicine. Databases and sources (see Table 3.1) can be used for such purposes including the following:

- dbSNP (Sherry et al., 2001): sequence variations and
- The 1000 Genomes Project (1000 Genomes Project Consortium et al., 2015): human genetic variations and diseases.

For instance, in the examination of SNP array data from the samples of esophageal cancer, bioinformatics segmentation algorithm was found to be useful (Bandla et al., 2012). The assessment revealed genomic variances with different frequencies from the data samples of both esophageal adenocarcinoma (EAC) and esophageal squamous cell carcinoma (ESCC). The study suggested that histology-specific therapeutic agents would be helpful for the different types of esophageal cancer.

In addition to the structure–function correlations, systems biology studies also emphasize the cellular pathways and interactions among different components. In recent years, more and more databases and sources are becoming available for such analysis (see Table 3.1), and here are some examples:

- Kyoto Encyclopedia of Genes and Genomes (KEGG) (Kanehisa et al., 2006): pathway maps for the interactions and networks about cellular processes and human diseases;
- Reactome (Croft et al., 2011): a pathway database for visualization and analysis;
- The Human Protein Reference Database (HPRD) (Keshava et al., 2009): a database of protein–protein interactions;
- IntAct: a database about molecular interactions;
- The MIPS Mammalian Protein–Protein Interaction Database (Pagel et al., 2005): about protein–protein interactions; and
- The Small Molecule Pathway Database (SMPDB) (Jewison et al., 2014): human small molecule pathways including metabolic, drug action, and metabolic disease pathways.

Furthermore, the genome-wide association studies (GWAS) and HTP technologies have been used widely for pharmacogenomics and other "omics" studies (e.g., proteomic, lipidomics, and metabolomics). Some examples of the "omics" resources (see Table 3.1) are:

- Gene Expression Omnibus (GEO) (Barrett and Edgar, 2006): gene expression profiles and functional genomics data;
- ArrayExpress: functional genomics data from HTP experiments;
- RNA-Seq Atlas (Krupp et al., 2012): gene expression profiles from next-generation sequencing;
- The Human Protein Atlas (HPA) (Uhlen et al., 2010): a tissue-based map for proteomic analysis;
- Cancer Proteomics database (Arntzen et al., 2015): a database about cancer proteomics;
- MitoMiner (Smith and Robinson, 2016): a database for mitochondrial proteomics;
- isoMETLIN (Cho et al., 2014): a database about isotope-based metabolomics;

- The Human Metabolome Database (HMDB) (Wishart et al., 2009): a database of small molecule metabolites for metabolomics and biomarker analyses;
- LipidHome (Foster et al., 2013): a database to support HTP mass spectrometry lipidomics;
- The LIPID MAPS Lipidomics Gateway (Fahy et al., 2007): a database of lipid-related genes and proteins for lipidomics studies; and
- Biomarkers and Systems Medicine (BSM, 2016): a collection of tools and databases about biomarkers to support the development of systems medicine.

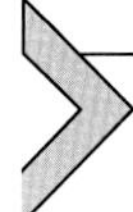

3.3 BIOINFORMATICS METHODS AND RESOURCES FOR EPIGENOMICS AND MICRORNA STUDIES

Recent development in epigenomics has shown that mechanisms such as DNA methylation may be critical for the elucidation of complex pathogenesis of the illnesses including cancer (Baek et al., 2013). Abnormal DNA methylation changes have been associated with various neurodevelopmental and neuropsychiatric disorders including schizophrenia and depression (Xin et al., 2012).

Many platforms and tools are becoming available for the disease-related epigenomics studies. Here are some examples (also see Table 3.2):

- DBCAT (DataBase of CpG Islands and Analytical Tools) (Kuo et al., 2011): a database for the analysis of DNA methylation profiles in human cancer;
- Cancer Methylome System (CMS) (Gu et al., 2013): a platform and viewer for differential methylation analysis about genome-wide methylation of tumors;
- MethylomeDB (Xin et al., 2012): a database of genome-wide brain DNA methylation profiles for the analyses of neuropsychiatric disorders;
- NGSmethDB (Hackenberg et al., 2011): a database and maps for studies of high-quality methylomes and differential methylation;
- EPITRANS (Cho et al., 2013): an epigenomics database about gene expression changes and epigenetic modification for the analyses of epigenome and transcriptome data; and
- EpiExplorer (Halachev et al., 2012): a web tool for analyzing epigenome data sets.

As small non-coding RNAs, microRNAs (miRNAs) have the key roles in various biological activities including energy and lipid metabolism.

Table 3.2 Bioinformatics Resources and Tools for Epigenomics and MicroRNA Studies

Tools	Web URLs	Contents
DBCAT	http://dbcat.cgm.ntu.edu.tw/	Methylation profiles in cancers
CMS	http://cbbiweb.uthscsa.edu/KMethylomes/	Cancer methylome
EpiExplorer	http://epiexplorer.mpi-inf.mpg.de/	Epigenomic analyses
EPITRANS	http://epitrans.org/EPITRANS/Service	Epigenetics and transcriptomics
MethylomeDB	http://www.neuroepigenomics.org/methylomedb/	Brain DNA methylation profiles
microRNA.org	http://www.microrna.org	MicroRNA expressions
miRBase	http://www.mirbase.org	MicroRNA database
miRDB	http://mirdb.org/miRDB	MicroRNA target predictions
miRGate	http://mirgate.bioinfo.cnio.es/miRGate/	Human, mouse, and rat microRNA–messenger RNA targets
miRNAMap 2.0	http://mirnamap.mbc.nctu.edu.tw	MicroRNA, genomic maps
miRò2	http://microrna.osumc.edu/miro/	The inference of microRNA associations
miRTarBase	http://mirtarbase.mbc.nctu.edu.tw	MicroRNA-target interactions
miRWalk2.0	http://zmf.umm.uni-heidelberg.de/apps/zmf/mirwalk2/	MicroRNA-target interactions
NGSmethDB	http://bioinfo2.ugr.es/NGSmethDB/index.php	DNA methylation

Studies of miRNAs would contribute to the better understanding of many complex disorders such as diabetes and cancer (Pescador et al., 2013). They are important messenger RNA (mRNA) regulators with unique disease expression signature profiles (Sandhu and Maddock, 2014). These features have indicated that miRNAs can be useful biomarkers.

Many bioinformatics databases and tools about miRNA are becoming available in recent years. Here are some examples of such resources (also see Table 3.2):

- The miRBase database (Kozomara and Griffiths-Jones, 2013): a searchable miRNA database containing annotations and miRNA sequences;
- miRGate (Andrés-León et al., 2015): human, mouse, and rat miRNA–mRNA targets;

- miRTarBase (Chou et al., 2016): a database about miRNA-target interactions that have been validated by experiments including microarray and next-generation sequencing;
- The microRNA.org site (Betel et al., 2008): a platform about miRNA target predictions and expression profiles;
- The miRNAMap 2.0 (Hsu et al., 2008): genomic maps of miRNAs and target genes in various species; and
- miRDB (Wong and Wang, 2015): a database and analytical tool for miRNA target prediction and functional annotations for various species.

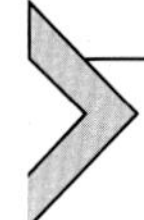

3.4 BIOINFORMATICS SUPPORT FOR THE STUDIES OF DISEASE PHENOTYPES AND DRUG RESPONSES

Based on the various "omics" studies, a pivotal investigation for translational bioinformatics and systems biology is to elucidate the genotype–phenotype and gene–drug correlations including symptoms and drug responses. Some resources helpful for such purposes are listed in the following (also see Table 3.3 and Fig. 3.1). More resources and methods about specific conditions and diseases such as inflammation, cardiovascular diseases, cancer, and aging can be found in Chapters 9–12.

- Online Mendelian Inheritance in Man (OMIM) (Hamosh et al., 2005): a database of human genes, phenotypes, and disorders;
- Gentrepid (George et al., 2006): a platform for candidate disease gene prediction;
- The Database of Genotypes and Phenotypes (dbGaP): a database about the genotype–phenotype interactions in humans;
- PhenoScanner (Staley et al., 2016): a database about human genotype–phenotype correlations;
- The Human Genome Epidemiology Network (HuGENet): a resource about public health genomics and epidemiology; and
- ClinicalTrials.gov: a database about clinical studies.

In addition, factors including nutrients and drugs are essential in understanding the genotype–phenotype interactions for personalized medicine. Some general resources are listed in the following (also see Table 3.3). More resources about drug discovery and development can be found in Chapter 8.

- National Health and Nutrition Examination Survey (NHANES): a program about the health and nutritional conditions including assessments from interviews and physical examinations;
- DrugBank (Wishart et al., 2006): a database about drugs and drug targets;

Table 3.3 Bioinformatics Resources and Tools for the Studies of Disease Phenotypes and Drug Responses

Tools	Web URLs	Contents
ChEMBL	https://www.ebi.ac.uk/chembl/	Drug-like small molecules
ClinicalTrials.gov	http://clinicaltrials.gov/	Clinical trials
dbGaP	http://www.ncbi.nlm.nih.gov/gap	Genotype–phenotype interactions
Drug Interaction Database (DIDB)	http://www.druginteraction-info.org/	Drug interactions
DrugBank	http://www.drugbank.ca	Drugs and targets
Drugs@FDA Database	https://www.accessdata.fda.gov/scripts/cder/drugsatfda/index.cfm	Drugs
FDA Adverse Event Reporting System (FAERS)	http://www.fda.gov/Drugs/GuidanceComplianceRegulatoryInformation/Surveillance/AdverseDrugEffects/default.htm	Adverse events and medication error reports
Gentrepid	http://www.gentrepid.org/	Genetic disorders
HuGENet	http://www.cdc.gov/genomics/hugenet/default.htm	Genetic variations
MedWatch	http://www.fda.gov/Safety/MedWatch/	FDA safety information
NHANES	http://www.cdc.gov/nchs/nhanes.htm	Health and nutritional status
Online Mendelian Inheritance in Man (OMIM)	http://www.ncbi.nlm.nih.gov/omim	Human genes and diseases
PhenoScanner	http://www.phenoscanner.medschl.cam.ac.uk/phenoscanner	Human genotype–phenotype correlations
SIDER (EMBL)	http://sideeffects.embl.de	Adverse drug reactions

- Drug Interaction Database (DIDB) program: a platform about drug development and drug–drug interactions;
- The Drugs@FDA: a platform about FDA-approved drugs that can be searched by drug names and ingredients;
- ChEMBL: a platform about drug ligands, compounds, targets, and assays;
- The FDA Adverse Event Reporting System (FAERS): a database about reported adverse events and medication errors associated with drugs and therapeutic biologic products;

- MedWatch: an FDA reporting program about safety and adverse events;
- SIDER: a resource about marketed medicines, adverse drug reactions, side effects, and drug targets.

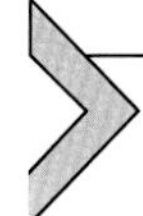

3.5 BIOINFORMATICS SUPPORT FOR THE SPATIOTEMPORAL STUDIES TOWARD DYNAMICAL MEDICINE

As discussed in Chapters 1 and 2, systems and dynamical studies across various spatial levels and temporal scales are fundamental for the development of systems and dynamical medicine. Translational bioinformatics methods are especially important for such efforts. As shown below, from molecular to cellular levels, from tissues to organs, many resources can be applied for studying genomic and proteomic dynamics (also see Table 3.4). More resources and methods about biomedical dynamics can be found in Chapter 7.

- Systems Science of Biological Dynamics database (SSBD) (Tohsato et al., 2016): a database of spatiotemporal dynamics in biology;
- Dynameomics (Van der Kamp et al., 2010): a database about molecular dynamics simulation including protein folding;
- The Kahn Dynamic Proteomics (Frenkel-Morgenstern et al., 2010): a resource about the dynamics of proteins in living human cells including the position and amounts;
- CHARMM-GUI (Jo et al., 2007): a platform for the simulations of molecular dynamics and mechanics;
- CellFinder (Stachelscheid et al., 2013): a platform about gene and protein expression profiles, phenotypes, development, and images associated with cell types;
- Arena3D (Secrier et al., 2012): a visualization resource about time-driven phenotypic variances including the networks and dynamic processes;
- MitoGenesisDB (Gelly et al., 2011): a database about the dynamics and biogenesis of mitochondrial protein formation;
- EUCLIS (EUCLock Information System) (Batista et al., 2007): a platform about circadian rhythms and chronobiology;
- The Allen Brain Atlas (Sunkin et al., 2013): a spatiotemporal data portal for studying the central nervous system (CNS) including gene expression data and cell types;
- The Brain Transcriptome Database (BrainTx) (Sato et al., 2008): a database about the visualization and analysis of transcriptome data about the stages and states of the brain;

Table 3.4 Bioinformatics Resources and Tools for Molecular Dynamics and Spatiotemporal Studies

Tools	Web URLs	Contents
Allen Brain Atlas	http://www.brain-map.org	Central nervous system spatiotemporal maps
Arena3D	http://arena3d.org	Time-driven phenotypic differences
CellFinder	http://cellfinder.org	Cell types in complex systems
CHARMM-GUI	http://www.charmm-gui.org/	Macromolecular dynamics
Dynameomics	http://www.dynameomics.org	Protein dynamics
EpiScanGIS	http://www.episcangis.org	Spatiotemporal clusters of diseases
EUCLIS	http://www.bioinfo.mpg.de/euclis/	Circadian systems biology
Eurexpress	http://www.eurexpress.org	Transcriptome in the mouse embryo
MitoGenesisDB	http://www.dsimb.inserm.fr/dsimb_tools/mitgene/	Spatiotemporal dynamics of mitochondria
Systems Science of Biological Dynamics (SSBD)	http://ssbd.qbic.riken.jp/	Quantitative data and microscopy images
The Brain Transcriptome Database (BrainTx)	http://www.cdtdb.neuroinf.jp/CDT/Top.jsp	Spatiotemporal gene expressions in mouse brains
The Kahn Dynamic Proteomics Database	http://www.weizmann.ac.il/mcb/UriAlon/DynamProt/	Protein dynamics
VectorMap	http://vectormap.si.edu/	Disease maps and distribution models for arthropod disease vector species

- The Eurexpress (Diez-Roux et al., 2011): a database about the transcriptome map of the mouse embryo;
- EpiScanGIS (Reinhardt et al., 2008): a surveillance and geographical information system in Germany about the timely information and distribution of meningococcal disease; and
- VectorMap (Kelly et al., 2015): a program that provides disease maps and distribution models about arthropod disease vector species.

3.6 CONCLUSION

In conclusion, the integration of bioinformatics and healthcare informatics is indispensable for the translation of systems biology and pharmacogenomics into personalized and systems dynamical medicine. These approaches are also the key to the identification of the spatiotemporal patterns such as the time-series assessments of the correlations between genetic structural variations and functional changes including drug responses.

It is essential to understand the cross talks among various systems levels and temporal scales including the genotype–phenotype correlations for finding systems-based biomarkers and for the classification of patient subgroups at different disease phases (see Chapters 1, 2, and 5). The systems-based and dynamical profiling may enable the discovery of more useful prognostic biomarkers and more effective preventive strategies.

Methods in translational bioinformatics such as data integration and data mining may empower the decision support efforts in both research and clinical settings (see Chapter 4). These approaches may improve the decision-making activities via more convenient communication and information access. They may promote KD and predictive modeling toward the optimal diagnosis and treatments.

REFERENCES

1000 Genomes Project Consortium, Auton, A., Brooks, L.D., Durbin, R.M., et al., 2015. A global reference for human genetic variation. Nature 526, 68–74. http://dx.doi.org/10.1038/nature15393.

Altschul, S.F., Gish, W., Miller, W., et al., 1990. Basic local alignment search tool. J. Mol. Biol. 215, 403–410.

Andrés-León, E., González Peña, D., Gómez-López, G., Pisano, D.G., 2015. miRGate: a curated database of human, mouse and rat miRNA-mRNA targets. Database (Oxford) 2015, bav035.

Arntzen, M.Ø., Boddie, P., Frick, R., Koehler, C.J., Thiede, B., 2015. Consolidation of proteomics data in the cancer proteomics database. Proteomics 15, 3765–3771.

Baek, S.-J., Yang, S., Kang, T.-W., et al., 2013. MENT: methylation and expression database of normal and tumor tissues. Gene 518, 194–200.

Bandla, S., Pennathur, A., Luketich, J.D., et al., 2012. Comparative genomics of esophageal adenocarcinoma and squamous cell carcinoma. Ann. Thorac. Surg. 93, 1101–1106.

Barrett, T., Edgar, R., 2006. Gene expression omnibus: microarray data storage, submission, retrieval, and analysis. Methods Enzymol. 411, 352–369.

Batista, R.T.B., Ramirez, D.B., Santos, R.D., et al., 2007. EUCLIS–an information system for circadian systems biology. IET Syst. Biol. 1, 266–273.

Betel, D., Wilson, M., Gabow, A., et al., 2008. The microRNA.org resource: targets and expression. Nucleic Acids Res. 36, D149–D153.

BSM, 2016. Biomarkers and Systems Medicine. http://pharmtao.com/health/category/systems-medicine/biomarkers-systems-medicine.

Cho, S.Y., Chai, J.C., Park, S.J., et al., 2013. EPITRANS: a database that integrates epigenome and transcriptome data. Mol. Cells 36, 472–475.
Cho, K., Mahieu, N., Ivanisevic, J., Uritboonthai, W., Chen, Y.-J., Siuzdak, G., Patti, G.J., 2014. isoMETLIN: a database for isotope-based metabolomics. Anal. Chem. 86, 9358–9361.
Chou, C.H., Chang, N.W., Shrestha, S., et al., 2016. miRTarBase 2016: updates to the experimentally validated miRNA-target interactions database. Nucleic Acids Res. 44, D239–D247. http://dx.doi.org/10.1093/nar/gkv1258.
Croft, D., O'Kelly, G., Wu, G., et al., 2011. Reactome: a database of reactions, pathways and biological processes. Nucleic Acids Res. 39, D691–D697.
Diez-Roux, G., Banfi, S., Sultan, M., et al., 2011. A high-resolution anatomical atlas of the transcriptome in the mouse embryo. PLoS Biol. 9, e1000582.
Fahy, E., Sud, M., Cotter, D., Subramaniam, S., 2007. LIPID MAPS online tools for lipid research. Nucleic Acids Res. 35, W606–W612.
Foster, J.M., Moreno, P., Fabregat, A., Hermjakob, H., Steinbeck, C., Apweiler, R., Wakelam, M.J.O., Vizcaíno, J.A., 2013. LipidHome: a database of theoretical lipids optimized for high throughput mass spectrometry lipidomics. PLoS One 8, e61951.
Frenkel-Morgenstern, M., Cohen, A.A., Geva-Zatorsky, N., et al., 2010. Dynamic proteomics: a database for dynamics and localizations of endogenous fluorescently-tagged proteins in living human cells. Nucleic Acids Res. 38, D508–D512.
Gelly, J.-C., Orgeur, M., Jacq, C., Lelandais, G., 2011. MitoGenesisDB: an expression data mining tool to explore spatio-temporal dynamics of mitochondrial biogenesis. Nucleic Acids Res. 39, D1079–D1084.
George, R.A., Liu, J.Y., Feng, L.L., et al., 2006. Analysis of protein sequence and interaction data for candidate disease gene prediction. Nucleic Acids Res. 34, e130.
Gu, F., Doderer, M.S., Huang, Y.-W., et al., 2013. CMS: a web-based system for visualization and analysis of genome-wide methylation data of human cancers. PLoS One 8, e60980.
Hackenberg, M., Barturen, G., Oliver, J.L., 2011. NGSmethDB: a database for next-generation sequencing single-cytosine-resolution DNA methylation data. Nucleic Acids Res. 39, D75–D79.
Halachev, K., Bast, H., Albrecht, F., et al., 2012. EpiExplorer: live exploration and global analysis of large epigenomic datasets. Genome Biol. 13, R96.
Halberg, F., Cornélissen, G., Katinas, G., et al., 2007. Chronomics and genetics. Scr. Medica (Brno) 80, 133–150.
Hamosh, A., Scott, A.F., Amberger, J.S., et al., 2005. Online Mendelian Inheritance in Man (OMIM), a knowledgebase of human genes and genetic disorders. Nucleic Acids Res. 33, D514–D517.
Hsu, S.-D., Chu, C.-H., Tsou, A.-P., et al., 2008. miRNAMap 2.0: genomic maps of microRNAs in metazoan genomes. Nucleic Acids Res. 36, D165–D169.
Jewison, T., Su, Y., Disfany, F.M., Liang, Y., Knox, C., Maciejewski, A., Poelzer, J., Huynh, J., Zhou, Y., Arndt, D., et al., 2014. SMPDB 2.0: big improvements to the small molecule pathway database. Nucleic Acids Res. 42, D478–D484.
Jo, S., Kim, T., Im, W., 2007. Automated builder and database of protein/membrane complexes for molecular dynamics simulations. PLoS One 2, e880.
Kanehisa, M., Goto, S., Hattori, M., et al., 2006. From genomics to chemical genomics: new developments in KEGG. Nucleic Acids Res. 34, D354–D357.
Kelly, D.J., Foley, D.H., Richards, A.L., 2015. A spatiotemporal database to track human scrub typhus using the VectorMap application. PLoS Negl. Trop. Dis. 9, e0004161.
Keshava Prasad, T.S., Goel, R., Kandasamy, K., et al., 2009. Human Protein Reference Database–2009 update. Nucleic Acids Res. 37, D767–D772.
Kozomara, A., Griffiths-Jones, S., 2013. miRBase: annotating high confidence microRNAs using deep sequencing data. Nucleic Acids Res.

Krupp, M., Marquardt, J.U., Sahin, U., et al., 2012. RNA-Seq Atlas – a reference database for gene expression profiling in normal tissue by next-generation sequencing. Bioinformatics 28, 1184–1185.
Kuo, H.-C., Lin, P.-Y., Chung, T.-C., et al., 2011. DBCAT: database of CpG islands and analytical tools for identifying comprehensive methylation profiles in cancer cells. J. Comput. Biol. 18, 1013–1017.
Larkin, M.A., Blackshields, G., Brown, N.P., et al., 2007. Clustal W and Clustal X version 2.0. Bioinformatics 23, 2947–2948.
Pagel, P., Kovac, S., Oesterheld, M., et al., 2005. The MIPS mammalian protein-protein interaction database. Bioinformatics 21, 832–834.
Pescador, N., Pérez-Barba, M., Ibarra, J.M., et al., 2013. Serum circulating microRNA profiling for identification of potential type 2 diabetes and obesity biomarkers. PLoS One 8, e77251.
Reinhardt, M., Elias, J., Albert, J., et al., 2008. EpiScanGIS: an online geographic surveillance system for meningococcal disease. Int. J. Health Geogr. 7, 33.
Rose, P.W., Beran, B., Bi, C., Bluhm, W.F., et al., 2011. The RCSB Protein Data Bank: redesigned web site and web services. Nucleic Acids Res. 39, D392–D401.
Sandhu, H., Maddock, H., 2014. Molecular basis of cancer-therapy-induced cardiotoxicity: introducing microRNA biomarkers for early assessment of subclinical myocardial injury. Clin. Sci. 126, 377–400.
Sato, A., Sekine, Y., Saruta, C., et al., 2008. Cerebellar development transcriptome database (CDT-DB): profiling of spatio-temporal gene expression during the postnatal development of mouse cerebellum. Neural Netw. 21, 1056–1069.
Secrier, M., Pavlopoulos, G.A., Aerts, J., Schneider, R., 2012. Arena3D: visualizing time-driven phenotypic differences in biological systems. BMC Bioinform. 13, 45.
Sherry, S.T., Ward, M.H., Kholodov, M., et al., 2001. dbSNP: the NCBI database of genetic variation. Nucleic Acids Res. 29, 308–311.
Sigrist, C.J.A., Cerutti, L., de Castro, E., et al., 2010. PROSITE, a protein domain database for functional characterization and annotation. Nucleic Acids Res. 38, D161–D166.
Smith, A.C., Robinson, A.J., 2016. MitoMiner v3.1, an update on the mitochondrial proteomics database. Nucleic Acids Res. 44, D1258–D1261.
Stachelscheid, H., Seltmann, S., Lekschas, F., et al., 2013. CellFinder: a cell data repository. Nucleic Acids Res. 42, D950–D958.
Staley, J.R., Blackshaw, J., Kamat, M.A., Ellis, S., Surendran, P., Sun, B.B., Paul, D.S., Freitag, D., Burgess, S., Danesh, J., et al., 2016. PhenoScanner: a database of human genotype-phenotype associations. Bioinformatics 32, 3207–3209.
Suh, K.S., Remache, Y.K., Patel, J.S., et al., 2009. Informatics-guided procurement of patient samples for biomarker discovery projects in cancer research. Cell Tissue Bank. 10, 43–48.
Sunkin, S.M., Ng, L., Lau, C., et al., 2013. Allen Brain Atlas: an integrated spatio-temporal portal for exploring the central nervous system. Nucleic Acids Res. 41, D996–D1008.
Tohsato, Y., Ho, K.H.L., Kyoda, K., Onami, S., 2016. SSBD: a database of quantitative data of spatiotemporal dynamics of biological phenomena. Bioinformatics.
Uhlen, M., Oksvold, P., Fagerberg, L., et al., 2010. Towards a knowledge-based human protein atlas. Nat. Biotechnol. 28, 1248–1250.
Van der Kamp, M.W., Schaeffer, R.D., Jonsson, A.L., et al., 2010. Dynameomics: a comprehensive database of protein dynamics. Structure 18, 423–435.
Wishart, D.S., Knox, C., Guo, A.C., et al., 2006. DrugBank: a comprehensive resource for in silico drug discovery and exploration. Nucleic Acids Res. 34, D668–D672.
Wishart, D.S., Knox, C., Guo, A.C., et al., 2009. HMDB: a knowledgebase for the human metabolome. Nucleic Acids Res. 37, D603–D610.
Wong, N., Wang, X., 2015. miRDB: an online resource for microRNA target prediction and functional annotations. Nucleic Acids Res. 43, D146–D152.

Xin, Y., Chanrion, B., O'Donnell, A.H., et al., 2012. MethylomeDB: a database of DNA methylation profiles of the brain. Nucleic Acids Res. 40, D1245–D1249.
Yan, Q., 2010. Translational bioinformatics and systems biology approaches for personalized medicine. Methods Mol. Biol. 662, 167–178.
Yan, Q., 2012. Translational bioinformatics in psychoneuroimmunology: methods and applications. Methods Mol. Biol. 934, 383–400.

CHAPTER FOUR

Data Integration, Data Mining, and Decision Support in Biomedical Informatics

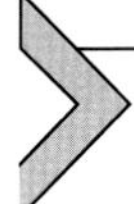

4.1 INTRODUCTION: DATA AND WORKFLOW INTEGRATION IN TRANSLATIONAL BIOINFORMATICS

In biomedical studies especially systems biology, a significant challenge comes from the heterogeneous data sets about the complex systems and relevant diseases (Dreher et al., 2012). The integration of the various data sets such as those from molecular, clinical, and imaging studies into coherent models may help elucidate the dynamic and heterogeneous properties of the complex diseases and validate robust biomarkers. Such integrative strategies would pave the ground toward the advancement of personalized medicine.

Many tools and approaches can be applied in translational bioinformatics to support the translation of scientific discoveries into better clinical results. As illustrated in Fig. 4.1, some of the important steps are data integration, data standardization, data mining, knowledge discovery (KD), and decision support. These methods are useful for both basic analyses, such as genomic studies, and clinical practice, such as electronic health records (EHRs).

The large volumes of data sets from both laboratory experiments and clinical trials may be valuable resources for the development of personalized and systems medicine. However, these data need to be well managed before they can be stored, shared, retrieved, and analyzed. Various reasons may complicate the data management processes, such as limited time and budgets, different resources and backgrounds, and knowledge domain barriers among researchers and clinicians.

To overcome the obstacles for more efficient data management in translational bioinformatics, strategies for data integration are essential (Brazhnik and Jones, 2007). Besides data integration, the diversity and heterogeneity in clinical and research environments demand workflow integration to enable

Translational Bioinformatics and Systems Biology Methods for Personalized Medicine
ISBN 978-0-12-804328-8
http://dx.doi.org/10.1016/B978-0-12-804328-8.00004-8

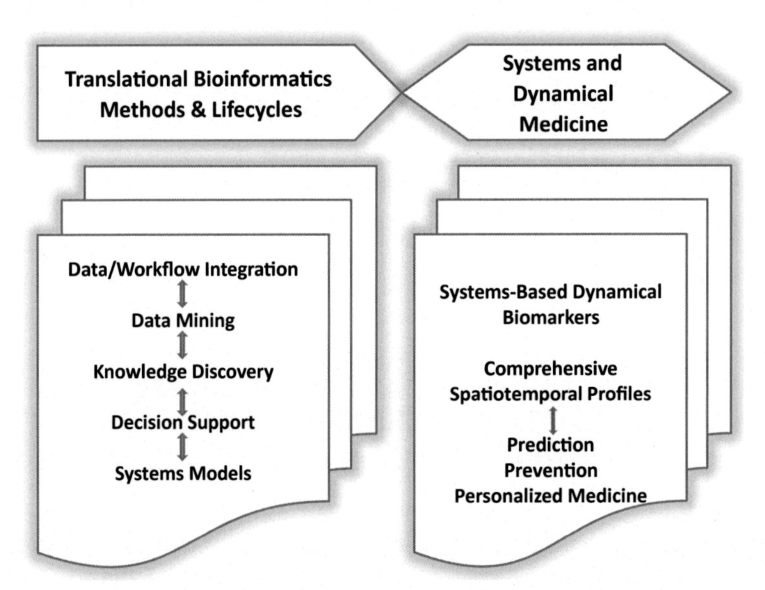

Figure 4.1 Translational bioinformatics methods for systems and dynamical medicine.

better decision support. Such steps are fundamental for further data mining and KD, especially for biomarker identifications and pattern recognitions to enable the construction of predictive models (Yan, 2010a). The better data exchanging and management may also help improve time and financial effectiveness.

The advantages of data integration have been demonstrated by various real-world applications. For example, a methodological integration framework was used to integrate data from different genome-wide sources and multiple tiers of biological regulation (Daemen et al., 2009). The data came from various "omics" studies including genomics, proteomics, transcriptomics, and epigenomics (see Chapter 3). The clinical data sets included samples from rectal cancer and prostate cancer patients. The integration processes enhanced the predictive power of clinical decision support models and allowed for better clinical outcomes. Such integrative strategies have demonstrated that they can be very useful for cost-efficient purposes and personalized therapies.

Such methods are critical for the translational research in systems biology. For instance, complex data from high-throughput (HTP) approaches, such as microarray tests, do not have any meaning without large-scale integration to identify the shared pathways in different cancer types (Dawany et al.,

2011). The gene names, sample sizes, and data types may be varied in different experiments and different studies. Such discrepancies may complicate the analyses including gene expressions in different cancer tissues. These difficulties make the data integration step indispensable.

An example of the solution was using data normalization methods for managing the microarray data from more than 80 laboratories with more than 4000 samples and more than 10 cancer tissue types (Dawany et al., 2011). The integration step resulted in the organized lists of genes for each cancer type as the potential biomarkers including various kinases and transcription factors.

Such real-world examples have demonstrated both usages and benefits of the data integration approaches. In another example of multiple sclerosis, data integration and systems biology strategies were found to enhance the efforts for biomarker discovery based on various system levels in the complex hierarchy of humans (Villoslada and Baranzini, 2012). More of such practical examples in various diseases can be found in Chapters 9–12.

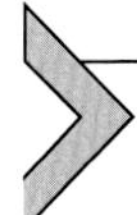

4.2 APPROACHES OF DATA AND WORKFLOW INTEGRATION

4.2.1 The Basic Data Integration Steps

As discussed previously, data integration can be applied in several steps starting from the recognition, gathering, and selection of various data sources (Yan, 2010a,b). These resources should meet the objectives from project requirement analyses. The data from various sources can then be collected, stored, cleaned, corrected, and updated for removing possible errors. Any inconsistencies in the data should also be fixed to organize the data. Common values should be combined. The representation formats or styles of the data can be changed and amended before they can be distributed and analyzed.

Most importantly, the data standardization is essential in all these steps including using the standards for data types, terms, names, and values. Some examples of commonly used standards for bioinformatics and healthcare informatics are available in Tables 4.1 and 4.2 of this chapter, as well as in Chapter 3.

These steps in data integration provide the transformational and evolutionary power toward KD and decision support in later stages. Biomedical data usually have problems including redundancies and discrepancies. These problems may lead to misperception and mistakes. The healthcare

Table 4.1 Bioinformatics Standards and Resources

Tools	Web URLs	Contents
Data Integration Platform for Systems Biology Collaborations (DIPSBC)	http://dipsbc.molgen.mpg.de/foswiki/DIPSBC/WebHome	Data integration for systems biology
Electronic Medical Records and Genomics (eMERGE)	https://www.mc.vanderbilt.edu/victr/dcc/projects/acc/index.php/Main_Page	Electronic medical records and genomics
Gene Ontology (GO)	http://www.geneontology.org/	Gene ontology
HUGO Gene Nomenclature Committee (HGNC)	http://www.genenames.org/	Gene nomenclature
Integrative Modeling Platform (IMP)	https://integrativemodeling.org/	Genome structural analyses via data integration
Personal Genome Project	http://www.personalgenomes.org/	Personal genomes
Systems Biology Markup Language (SBML)	http://sbml.org/Main_Page	Biological process modeling
NCI Center for Biomedical Informatics and Information Technology (CBIIT)	https://cbiit.nci.nih.gov/	Cancer studies

and bioinformatics standards can be applied as references to clean up such errors. For example, the genetic nomenclature source Gene Ontology (GO) can be used for unifying gene names (see Table 4.1).

In addition, the combination of different data from various resources including databases and software tools can be one of the first steps in data integration. A practical strategy is using a shared web-based infrastructure to gather and link data and information, such as databases, data warehouses, portals, and various integrative frameworks.

Recent techniques such as XML- and Wiki-based data integration methods may also facilitate the efforts for indexing, querying, and mining of heterogeneous data sets (Dreher et al., 2012). The cloud computing technology may provide powerful support for storing and accessing data. These approaches can facilitate better data sharing among different knowledge domains and different physical locations. The integrative platforms can also improve the workflow efficiencies in both laboratory and clinical departments with the potential benefits of saving costs.

Table 4.2 Health Informatics Standards and Resources

Tools	Web URLs	Contents
Digital Imaging and Communications in Medicine (DICOM)	http://medical.nema.org/	Medical imaging data standards
Genetic Information Nondiscrimination Act (GINA)	http://www.genome.gov/24519851	Genetic information nondiscrimination act
The Health Insurance Portability and Accountability Act of 1996 (HIPAA) and the Patient Safety and Quality Improvement Act of 2005 (PSQIA)	http://www.hhs.gov/ocr/privacy/	Patient privacy, safety, security rules
Health Level Seven International (HL7)	http://www.hl7.org/	Health information systems interoperability
International Classification of Diseases (ICD)	http://www.who.int/classifications/icd/en/	Disease classifications
Logical Observation Identifiers Names and Codes (LOINC)	http://loinc.org/	Laboratory observations
Medical Dictionary for Regulatory Activities (MedDRA)	http://www.meddra.org/	Medical terminology for adverse events
RxNorm	https://www.nlm.nih.gov/research/umls/rxnorm/	Drug terminologies
Systematized Nomenclature of Medicine–Clinical Terms (SNOMED-CT)	http://www.nlm.nih.gov/research/umls/Snomed/snomed_main.html	Clinical terminology
Unified Medical Language System (UMLS)	http://www.nlm.nih.gov/research/umls/	Medical terminology integration

For instance, to support the research about the gene expression patterns of the fruit fly *Drosophila melanogaster*, an integrative platform was constructed for functional genomics analyses (Miles et al., 2010). The effort merged Semantic Web Standards and Web 2.0 design patterns to facilitate more convenient access and searching functions for the comparison of the expression data from various sources. Ontology and data mapping strategies were critical in the data integration processes.

In addition to genomics data, data integration approaches have also been proven useful in the management of tissue banking data from biospecimens in different groups (Amin et al., 2010). Translational bioinformatics methods were applied to support the studies of cancer by using standardized annotations and data modeling for better data access from a wide range of clinical information systems.

An integrative platform was built to support queries about organ-specific biorepositories (Amin et al., 2010). The data architecture combined various resources such as the Specialized Programs of Research Excellence (SPORE) Head and Neck Neoplasm Database. Such integrative framework demonstrated the usefulness of bioinformatics strategies to support more effective translational efforts and collaboration.

As another example about depression, systems biology integration of proteomic data and functional enrichment analyses showed the importance of the immune, neurotrophic, and glutamatergic signaling pathways (Carboni et al., 2016). Methods of data integration included the network analysis of the proteomic data. Various resources were used for the data integration processes, including the databases of IntAct, HRPD, and GO for biological terms and pathways (see Chapter 3 and the next section of this chapter). More of such data integration examples in different diseases can be found in Chapters 9–12.

4.2.2 Bioinformatics and Health Informatics Resources for Standardization

For more effective decision support in both clinical and experimental environment involving various knowledge domains, standardization is the key solution for interoperability problems. Both bioinformatics and healthcare informatics resources are needed for the standardization step in translational medicine, as some of the examples listed in Tables 4.1 and 4.2 and in Chapter 3. Such standards are pivotal for the development of EHRs and clinical decision support systems (CDSSs).

Here are some examples that can be applied as the genetic nomenclature references (also see Table 4.1):

- GO: the ontology, concepts, and annotations of genes and functions;
- HUGO Gene Nomenclature Committee (HGNC): symbols and names of genes; and
- NCI Center for Biomedical Informatics and Information Technology (CBIIT): a program to support studies in cancer informatics.

Here are some example systems that can be applied as the clinical terminology references (also see Table 4.2):

- Systematized Nomenclature of Medicine—Clinical Terms (SNOMED CT): standards about clinical health information to solve interoperability issues;
- International Classification of Diseases (ICD): a standard about diseases and disease patterns for clinical data and billing;
- Logical Observation Identifiers Names and Codes (LOINC): a code system with universal identifiers for organizing laboratory data from tests, measurements, and observations;
- Digital Imaging and Communications in Medicine (DICOM): the standard for clinical imaging data;
- Unified Medical Language System (UMLS): a reference platform for the integration of terminologies, classifications, and coding standards;
- Health Level Seven International (HL7): a platform for the integration and sharing of electronic health information for medical practice;
- Medical Dictionary for Regulatory Activities (MedDRA): a platform about standardized medical terminology and regulatory information for medical products; and
- RxNorm: a system about drug vocabularies and normalized names for clinical drugs used in drug management and interaction programs.

4.2.3 The Integration of Biological and Medical Informatics

To meet the goals of translational bioinformatics, the integration of biological and medical informatics would be needed (see Chapter 1). Such combination would facilitate the sharing of both genomics and clinical data and address the genotype–phenotype correlations to support personalized medicine. Some efforts have already been made for such purposes (see Table 4.1), such as:

- eMERGE (Electronic Medical Records and Genomics): a combination of DNA information with electronic medical record (EMR) systems for genomic medicine;
- Personal Genome Project: an effort to share genomic, health, and trait data;
- Systems Biology Markup Language (SBML): a platform for bioinformatics modeling of biological processes including metabolism and signaling pathways;
- Data Integration Platform for Systems Biology Collaborations (DIPSBC) (Dreher et al., 2012): data integration for systems biology; and

- Integrative Modeling Platform (IMP) (Baù and Marti-Renom, 2012): genome structural analyses via data integration.

Furthermore, such integration of bioinformatics and healthcare informatics in CDSSs may face ethical, legal, privacy, and societal issues that should be addressed. Mutual efforts and cooperation from the societies of both bioinformatics and healthcare informatics are necessary. Here are some examples of the available efforts (see Table 4.2):

- Genetic Information Nondiscrimination Act (GINA): to protect people from discrimination based on genetic information, including health coverage and employment;
- The Health Insurance Portability and Accountability Act of 1996 (HIPAA) privacy and security rules and the Patient Safety and Quality Improvement Act of 2005 (PSQIA) patient safety rule: rights and rules about health information, which need to be complied by biomedical informatics systems including EHRs.

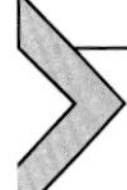

4.3 DATA MINING AND KNOWLEDGE DISCOVERY IN TRANSLATIONAL BIOINFORMATICS

As discussed previously, the data integration processes including cleaning, correcting, and standardization can transform unorganized data into usable information. Based on such data management, data mining approaches can be applied for KD such as finding patterns and correlations. This is a comprehensive process and a life cycle starting from requirement and domain analyses, through data integration, management, and data mining to support better decision making (see Fig. 4.1).

This is an iterative and interactive cycle that feedback from each step can be used to make improvements in the previous steps of the evolving life cycle. The ultimate goal is to promote decision support and knowledge representation to build predictive systems biology models that can be translated into better diagnosis and interventions in systems medicine (Yan, 2010a,b; also see Chapter 1).

Many real-world examples have proven the effectiveness of such life cycles in translational bioinformatics. For example, the KD methods were used in analyzing the reports from the World Health Organization (WHO) Adverse Drug Reaction (ADR) database (Bate et al., 2008). The informatics approaches were found useful for addressing the unpredicted ADRs for further review and for the examination of complex data sets from patient records.

In another example, a web-based integrative platform was constructed for the analysis of gene expression profiles with searching tools (Krupp et al., 2012). The comprehensive platform has connections to other resources, microarray profiles, signaling networks, and gene ontologies. These functions can facilitate data mining efforts such as comparing tissue-specific expression profiles to identify patterns.

The correlations between genotypic changes and tissue phenotypes are good examples for the across-level analyses in systems biology. Furthermore, semantic web technologies can be utilized to manage pharmacogenomics discoveries to support drug discovery and medical decision making as the essential function of CDSSs (Samwald et al., 2012).

Various data mining approaches are available for translational bioinformatics, such as data clustering, decision trees, Bayesian networks, genetic algorithms, and artificial neural networks (ANNs) (Yan, 2010a,b). Such strategies can help identify biomarkers, correlations, and patterns to support diagnosis, drug design, and treatment selections.

For instance, decision trees can be used for the investigation of protein–protein interactions in systems biology (Vallabhajosyula and Raval, 2010). Supervised classification methods have been found helpful for assessing microarray expression profiles for functional analyses and categorizing proteins to support predictive models (Brun et al., 2003).

Considering the conditions of inflammation, the mechanism underlying many diseases, the applications of the data modeling methods would be helpful (see Chapter 9). For instance, dynamic mathematical modeling methods such as agent-based modeling (ABM) and equation-based modeling (EBM) have been suggested (Vodovotz and An, 2010). Immunoinformatics programs can also be utilized for the investigations of the structure–function correlations of the immune system (Yan, 2010c).

Statistical analysis methods such as Bayesian networks have also been useful. In an evaluation of gene expression data in various time series, Gaussian Bayesian networks were applied together with reverse engineering regulatory networks for assessing the multidimensional data for building systems biology models (Grzegorczyk, 2010).

For better querying, knowledge representation, and pattern recognition, approaches including text mining and natural language processing (NLP) may be applied (Cohen and Hersh, 2005). NLP strategies have been found especially helpful for clinical narratives, text preprocessing, named entity recognition (NER), as well as the abstraction and finding of correlations

(Demner-Fushman et al., 2009). These can be applied for decision support in translational efforts in a wide range of areas such as scientific, administrative, and social issues.

To support dynamical assessments including chronobiology studies, data mining strategies can be applied to identify spatiotemporal patterns and correlations to build dependency models (Lopes et al., 2013). For instance, ABM has been found helpful for modeling nonlinear complexity across different system levels from cells to societies (Kaul and Ventikos, 2013).

In an examination of the circadian-associated gene expression patterns of the lung transcriptome, various data mining and translational bioinformatics methods were used including BLAST, MATLAB, and clustering algorithms (Sukumaran et al., 2011). The rhythmic patterns discovered using these strategies may contribute to systemic understanding of the dynamics of lung diseases and drug responses.

In the studies of breast cancer, ANN was applied for the dynamic thermal analysis (Salhab et al., 2006). An agent-based model was found useful for detecting mammary ductal epithelium dynamics to support the better understanding of both molecular and cellular pathogenesis (Chapa et al., 2013). More of such examples are available in Chapters 9–12.

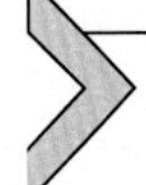

4.4 CONCLUSION: DECISION SUPPORT IN TRANSLATIONAL BIOINFORMATICS

As discussed previously, an important task in translational bioinformatics and systems biology is the investigation of the genotype–phenotype correlations by integrating genomic data with EMRs. Such multilevel strategies based on the information of genes, cells, drugs, interactions, and functional annotations are critical for decision support toward better diagnosis, prognosis, and personalized interventions (Peleg and Tu, 2006).

The term "decision support" refers to a broad meaning from data management to data mining, from diagnostic decisions to patient-centered clinical options (Peleg and Tu, 2006). Decision support is the key element of translational bioinformatics, from better data access to communication, from better drug design to treatment recommendations. It is the necessary step to bring the "right knowledge to the right people in the right form at the right time" (Yan, 2010a,b,c).

The integration and modeling of data and workflow are critical for effective decision support. Such purposes can also be achieved with the application of a standard object-oriented modeling tool Unified Modeling Language (UML) (Peleg and Tu, 2006; Yan, 2010b). UML can be especially useful in the life cycle for CDSS requirement analysis, system design, and implementation.

A practical example is the modeling of the clinical trial workflow using UML. The applications of UML diagrams, especially the activity diagram, were found helpful for modeling the clinical trials domain to support international sites in rheumatology and oncology for comparative investigations (de Carvalho et al., 2010). Such modeling strategies allowed for the translational processes of moving the paper notes to digital records to facilitate both nursing and administrative procedures.

In conclusion, the strategies of real-time data collection, mining, and predictive modeling would enable the integration of the health monitoring data elements with the EHRs to support the better applications by healthcare providers (Shameer et al., 2016). The combination of population-scale biomedical and clinical data may allow for grouping patients for proactive health promotion and tracking the disease trajectories. These are critical elements in clinical decision making, wellness care, and personalized medicine.

REFERENCES

Amin, W., Singh, H., Pople, A.K., Winters, S., Dhir, R., Parwani, A.V., Becich, M.J., 2010. A decade of experience in the development and implementation of tissue banking informatics tools for intra and inter-institutional translational research. J. Pathol. Inform. 1.

Bate, A., Lindquist, M., Edwards, I.R., 2008. The application of knowledge discovery in databases to post-marketing drug safety: example of the WHO database. Fundam. Clin. Pharmacol. 22 (2), 127–140.

Baù, D., Marti-Renom, M.A., 2012. Genome structure determination via 3C-based data integration by the Integrative Modeling Platform. Methods 58, 300–306.

Brazhnik, O., Jones, J.F., 2007. Anatomy of data integration. J. Biomed. Inform. 40 (3), 252–269.

Brun, C., Chevenet, F., Martin, D., Wojcik, J., Guénoche, A., Jacq, B., 2003. Functional classification of proteins for the prediction of cellular function from a protein-protein interaction network. Genome Biol. 5 (1), R6.

Carboni, L., Nguyen, T.-P., Caberlotto, L., 2016. Systems biology integration of proteomic data in rodent models of depression reveals involvement of the immune response and glutamatergic signaling. Proteom. Clin. Appl. 10 (12), 1254–1263.

de Carvalho, E.C.A., Jayanti, M.K., Batilana, A.P., Kozan, A.M.O., Rodrigues, M.J., Shah, J., Loures, M.R., et al., 2010. Standardizing clinical trials workflow representation in UML for international site comparison. PLoS One 5 (11), e13893.

Chapa, J., Bourgo, R.J., Greene, G.L., et al., 2013. Examining the pathogenesis of breast cancer using a novel agent-based model of mammary ductal epithelium dynamics. PLoS One 8, e64091.

Cohen, A.M., Hersh, W.R., 2005. A survey of current work in biomedical text mining. Brief. Bioinform. 6 (1), 57–71.

Daemen, A., Gevaert, O., Ojeda, F., Debucquoy, A., Suykens, J.A., Sempoux, C., Machiels, J.-P., et al., 2009. A kernel-based integration of genome-wide data for clinical decision support. Genome Med. 1 (4), 39.

Dawany, N.B., Dampier, W.N., Tozeren, A., 2011. Large-scale integration of microarray data reveals genes and pathways common to multiple cancer types. Int. J. Cancer (J. Int. Du Cancer) 128 (12), 2881–2891.

Demner-Fushman, D., Chapman, W.W., McDonald, C.J., 2009. What can natural language processing do for clinical decision support? J. Biomed. Inform. 42 (5), 760–772.

Dreher, F., Kreitler, T., Hardt, C., Kamburov, A., Yildirimman, R., Schellander, K., Lehrach, H., Lange, B.M.H., Herwig, R., 2012. DIPSBC–data integration platform for systems biology collaborations. BMC Bioinform. 13, 85.

Grzegorczyk, M., 2010. An introduction to Gaussian Bayesian networks. Methods Mol. Biol. (Clifton, NJ) 662, 121–147.

Kaul, H., Ventikos, Y., 2013. Investigating biocomplexity through the agent-based paradigm. Brief. Bioinform.

Krupp, M., Marquardt, J.U., Sahin, U., et al., 2012. RNA-Seq Atlas–a reference database for gene expression profiling in normal tissue by next-generation sequencing. Bioinformatics 28, 1184–1185.

Lopes Rda, S., Resende, N.M., Honorio-França, A.C., França, E.L., 2013. Application of bioinformatics in chronobiology research. ScientificWorldJournal 2013, 153839.

Miles, A., Zhao, J., Klyne, G., White-Cooper, H., Shotton, D., 2010. OpenFlyData: an exemplar data web integrating gene expression data on the fruit fly *Drosophila melanogaster*. J. Biomed. Inform. 43 (5), 752–761.

Peleg, M., Tu, S., 2006. Decision support, knowledge representation and management in medicine. Yearb. Med. Inform. 72–80.

Salhab, M., Keith, L.G., Laguens, M., et al., 2006. The potential role of dynamic thermal analysis in breast cancer detection. Int. Semin. Surg. Oncol. 3, 8.

Samwald, M., Coulet, A., Huerga, I., et al., 2012. Semantically enabling pharmacogenomic data for the realization of personalized medicine. Pharmacogenomics 13, 201–212.

Shameer, K., Badgeley, M.A., Miotto, R., Glicksberg, B.S., Morgan, J.W., Dudley, J.T., 2016. Translational bioinformatics in the era of real-time biomedical, health care and wellness data streams. Brief. Bioinform.

Sukumaran, S., Jusko, W.J., Dubois, D.C., Almon, R.R., 2011. Light-dark oscillations in the lung transcriptome: implications for lung homeostasis, repair, metabolism, disease, and drug action. J. Appl. Physiol. 110, 1732–1747.

Vallabhajosyula, R.R., Raval, A., 2010. Computational modeling in systems biology. Methods Mol. Biol. (Clifton, NJ) 662, 97–120.

Villoslada, P., Baranzini, S., 2012. Data integration and systems biology approaches for biomarker discovery: challenges and opportunities for multiple sclerosis. J. Neuroimmunol. 248, 58–65.

Vodovotz, Y., An, G., 2010. Systems biology and inflammation. Methods Mol. Biol. (Clifton, NJ) 662, 181–201.

Yan, Q., 2010a. Translational bioinformatics and systems biology approaches for personalized medicine. Methods Mol. Biol. (Clifton, NJ) 662, 167–178.

Yan, Q., 2010b. Bioinformatics for transporter pharmacogenomics and systems biology: data integration and modeling with UML. Methods Mol. Biol. (Clifton, NJ) 637, 23–45.

Yan, Q., 2010c. Immunoinformatics and systems biology methods for personalized medicine. Methods Mol. Biol. (Clifton, NJ) 662, 203–220.

PART TWO

Applications in Basic Sciences

CHAPTER FIVE

Applying Translational Bioinformatics for Biomarker Discovery

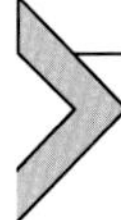

5.1 INTRODUCTION: CONCEPTS AND APPROACHES

5.1.1 The Basic Concepts and Types of Biomarkers

The discovery of biomarkers has become one of the major tasks in biomedical research. The identification of effective biomarkers has been suggested as one of the biggest challenges but also the essential factor for the successful rates of translational medicine (Wehling, 2008; Fu et al., 2010). In recent years, the finding of predictive and prognostic biomarkers has become one of the quickest developing areas.

An essential component of personalized medicine is useful biomarkers for quantified and more precise diagnosis and prognosis. They are also pivotal for discovering effective drug targets and choosing the optimal interventions. Biomarkers are objectively measured and assessed indicators of normal physiological activities, pathological conditions, as well as treatment responses (Yan, 2011). Biomarkers such as those involved in metabolomics can be tested in various samples including saliva, blood, and urine.

Biomarkers have been proven useful for the investigation of various illnesses including cancer, cardiovascular diseases, neurological diseases, respiratory diseases, and autoimmune diseases (Aich et al., 2009; Emerson et al., 2009; Chia et al., 2008; Knudsen et al., 2008; Ozkisacik et al., 2006). For example, in atherosclerotic disease, the novel local plaque biomarkers were applied for supporting the prediction of adverse reactions, the classification of patient subgroups, and personalized choices for more effective interventions (Hurks et al., 2009).

Biomarkers can be used to characterize the attributes of diseases. Dynamical biomarkers may describe the conditions of the disease during certain periods at specific time points. They may also be applied to track the stages and severities of diseases during their progression. In addition to diagnosis, prognosis, disease profiling, and classification, biomarkers can be

Translational Bioinformatics and Systems Biology Methods for Personalized Medicine
ISBN 978-0-12-804328-8
http://dx.doi.org/10.1016/B978-0-12-804328-8.00005-X

very useful for early disease detections for the prediction and prevention of disease risks. They may have pivotal roles in almost every step of translational medicine, from research design to drug discovery and development, from optimized treatment to outcome assessment.

Biomarkers have been grouped according to their applications (Biomarkers Definitions Working Group, 2001; Vasan, 2006; Markman and Tabernero, 2013). As summarized in Table 5.1, "antecedent biomarkers" are usually used to recognize the risks of disease development. The inexpensive "screening biomarkers" can be applied in the screening for subclinical disorders and prediction with high sensitivity and specificity.

The tests for "diagnostic biomarkers" need to be easy to run in clinical practice to make them useful in the identification and diagnosis of illnesses. To describe and classify the severities of different illnesses, the "staging biomarkers" can be used. To detect the disease recurrence and the therapeutic responses, and to predict the disease development and progression, "prognostic biomarkers" can be used. With high sensitivity and specificity, these biomarkers may reflect intraindividual variations and clinical outcomes for the assessment of treatment efficacy.

Table 5.1 A Summary of Different Types of Biomarkers

Biomarker Application Types	Applications and Features
Antecedent biomarkers	Identification of the disease risks
As surrogate endpoints	Prediction of the clinical benefits, outcomes, and therapeutic effectiveness
Diagnostic biomarkers	Disease diagnosis and easy to run in the clinic
PD biomarkers	Indications of drug effects, mechanisms of action (MOA), and optimal biologic dose (OBD)
Predictive biomarkers	Prediction of the therapeutic efficacy for the selection of personalized treatment to avoid adverse events
Prognostic biomarkers	Prediction of the disease progression and therapeutic responses with high sensitivity and specificity, highlighting intraindividual variations and clinical outcomes
Resistance biomarkers	Identification of primary and acquired drug resistance
Screening biomarkers	Screening for subclinical illnesses with high sensitivity and specificity, useful for prediction, usually inexpensive
Staging biomarkers	For classifying the disease severity

Certain biomarkers may serve as surrogate endpoints for the prediction of clinical advantages, outcomes, safety, and therapeutic effectiveness. These biomarkers can be identified based on epidemiological, treatment, pathophysiological, or other clinical evidences. Their applications may replace the clinical endpoints. Here the clinical endpoints refer to the variables that manifest the feelings, functions, or outcomes of patients (Biomarkers Definitions Working Group, 2001).

In addition, "pharmacodynamic (PD) biomarkers" may represent the drug effects such as molecular and functional influences before and after an intervention to identify the changes from the baseline. They can be very useful for the examination of molecular and cellular effects, as well as the mechanisms of action (MOA) of the drugs. They may also be applied for the pharmacokinetics (PK)/PD modeling of dosages and schedules to achieve the optimal biologic dose (OBD).

For the practice of personalized medicine, the "predictive biomarkers" can be especially helpful to predict the therapeutic efficacy to improve the selection of individualized treatment, to reduce adverse reactions, and to accomplish the optimized results. The type of "resistance biomarkers" may be used to indicate both primary and acquired drug resistance (see Table 5.1).

5.1.2 Steps and Pipelines of Biomarker Discovery

For biomarker discovery, a series of steps may be applied (Chau et al., 2008; Vilar and Tabernero, 2013; Markman and Tabernero, 2013; Biomarkers Definitions Working Group, 2001). These steps include:

- discovery,
- qualification,
- verification,
- research assay optimization,
- clinical validation, and
- commercialization.

Specifically, the "qualification" step is an integrative process for mapping a biomarker to a biological state with the setup of a surrogate endpoint exhibiting the linkages between the last steps of the discovery and the verification steps. The "validation" steps emphasize the assay optimization and clinical validation. These steps may be used to assess the operations of a test for the optimal analytical processes to support the formation of precise and reproducible biomarkers for clinical utilizations.

To support systems biology–based studies, different levels of biomarkers can be detected using different techniques (Markman and Tabernero, 2013). For example, sequencing techniques, gene expression studies, and high-throughput (HTP) microarrays can be applied for biomarkers at the molecular level including DNAs and RNAs.

Proteomics, crystallography, and immunohistochemistry techniques can be used to detect proteins and peptides as biomarkers (Markman and Tabernero, 2013). Cytogenetics approaches can be used to examine chromosomes. For the detection of physiological composite endpoints, imaging techniques can be applied including positron emission tomography (PET) scans and dynamic magnetic resonance imaging.

5.1.3 The Clinical Values of Biomarkers

To make biomarkers useful to be applied in clinical environments, some requirements need to be met including the reliability and measurability. Useful biomarkers should be specific and sensitive with predicative properties for the stratification of disease risks and classification of patient subgroups.

To support personalized and systems medicine (see Chapter 1), the identification of accurate and robust biomarkers is necessary for the early diagnosis and prognosis of complex diseases. Many factors can affect the clinical values of biomarkers (Vasan, 2006; Deeks and Altman, 2004), including:

- simplicity of analyses: so that they can be used by clinicians,
- sensitivity: the capacity of a test to identify specific conditions when they are really existed,
- specificity: the capacity of a test to exclude the specific conditions in those who do not have the illness,
- accuracy: having high sensitivity and specificity at certain cut points,
- reproducibility: the assay results are repeatable and reproducible,
- predictive value: the capacity of a test for the prediction of the true positives or true negatives as expected,
- interpretative capacity: the capacity to explain the clinical outcomes, and
- the likelihood ratio: the integration of information about the sensitivity and specificity, and the indication of the influence of the positive or negative results on the likelihood of having the disease.

Furthermore, the properties of complexity in diseases usually cannot be fully represented by one single biomarker or a few isolated biomarkers. To elucidate the comprehensive dysfunctions in the regulatory pathways and networks, systems-based biomarkers identified using multiparameter examinations are needed to assist more precise and effective diagnosis, prognosis, and interventions.

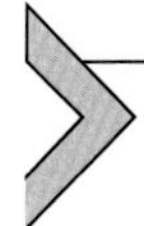

5.2 CHALLENGES AND TRANSLATIONAL BIOINFORMATICS METHODS FOR BIOMARKER DISCOVERY

As discussed earlier, biomarkers are critical for diagnosis, prognosis, disease profiling, drug discovery, and outcome assessment in personalized medicine. However, the pathological and treatment complexity as well as the dynamical changes in diseases make it very challenging to discover and validate useful biomarkers. Single and isolated biomolecules are insufficient to illustrate the composite biomedical functions as multiple interactions should be considered in the same disease phenotype (see Chapters 1 and 2).

Conventional approaches including symptom checklists have been ineffective for identifying appropriate biomarkers to characterize the multifaceted pathophysiological conditions, disease phases, and disease heterogeneity. Too much complexity and unknown may be the reasons for inefficient "trial-and-error" endeavors, time-consuming and high-cost experiments for identifying new drugs (Wierling et al., 2015).

For instance, although pathology type immunohistochemical markers are commonly utilized, their basic mechanisms and clinical indications are still unclear to be thoroughly applied in clinical pathology (Abu-Asab et al., 2011). In addition to representing the usual properties of an illness, biomarkers should also be able to discern the disease subtypes and the ranges of the disease heterogeneity.

To overcome the conceptual and practical obstacles in biomarker identification, systems biology methods incorporating translational bioinformatics and various "omics" strategies may help (see Chapter 3). Approaches such as data modeling may improve the systemic profiling of disease subtypes. As illustrated in Fig. 5.1, such approaches may address the complex factors at different systems levels from genetic polymorphisms to etiological diversities and environmental impacts (also see Chapters 2 and 7).

The systems biology-based biomarkers may be defined by focusing on the interactions among molecules, cells, tissues, organs, populations, and the micro- and macroenvironment (see Fig. 5.1). Biomarkers with dynamical features may help differentiate normal states from diseases during different phases and the divergences among individuals for patient classifications and subgroups.

Systems-based biomarkers may improve the selection and validation of drug targets and PD studies (Day et al., 2009). In addition to genetic biomarkers, other indicators beyond the molecular level such as imaging markers should also be included to characterize the overall pathophysiological

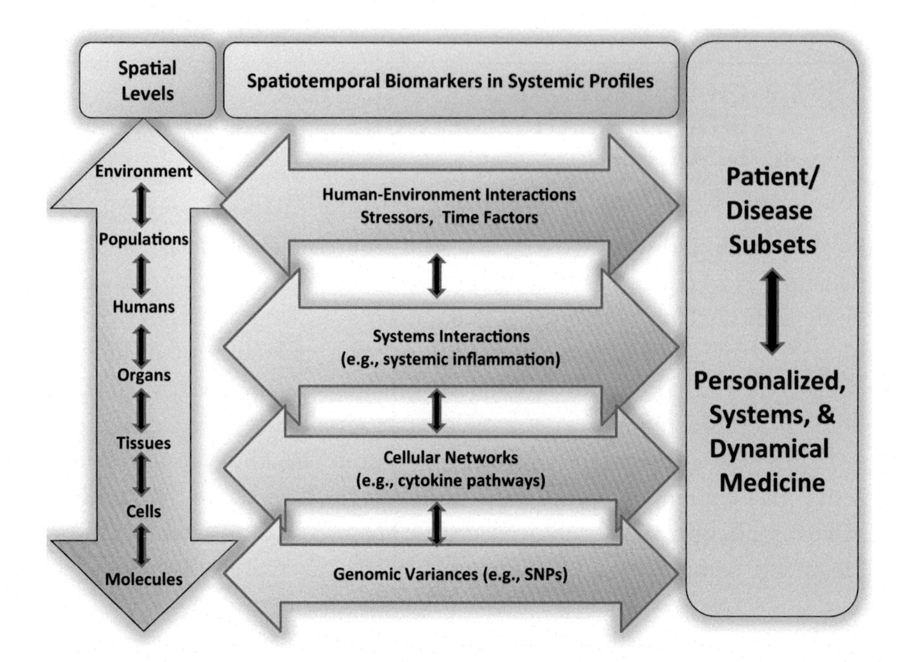

Figure 5.1 Biomarkers at different systems levels for systems and dynamical medicine.

conditions (Aich et al., 2009). Biomarkers based on cellular pathway-driven assessments have been suggested to be useful for monitoring disease activities such as those in cancer and cardiovascular disease (Azuaje et al., 2011).

Data integration is critical in biomarker discovery (see Chapter 4). For example, in the integration and analyses of microRNA (miRNA) databases such as miRWalk (see Chapter 3) and using combinatorial target prediction algorithms, some potential urinary exosomal miRNA biomarkers were identified for diabetic nephropathy (Eissa et al., 2016). The integrative bioinformatics methods have been found to be successful for detecting potential prognostic biomarkers for clear cell renal cell carcinoma. With the comprehensive pathway analysis, the modeling of interaction networks, and the integration of publicly available data sets including those from microRNA and protein expression profiles, three factors were revealed as possible biomarkers including AHR, GRHL2, and KIAA0101 (Butz et al., 2014).

The applications of data integration have been proven to be very useful for finding potential biomarkers for bortezomib resistance in multiple myeloma (Fall et al., 2014). The integration of large patient databases, the profiling of whole transcriptome, and the establishment of laboratory-based models may revolutionize the processes for identifying rational drug targets tracking therapeutic responses. By applying bioinformatics tools for the assessment of nuclear magnetic resonance (NMR) metabolomics, diagnostic, prognostic, and predictive biomarkers may be discovered for the translation into early preventions and interventions (Puchades-Carrasco et al., 2016).

Decision support in biomarker discovery is one of the key tasks in translational bioinformatics (Azuaje et al., 2011; also see Chapter 4). Such strategies would enable the integration and mining of large and diversified data sets for finding novel associations to improve the predictive reproducibility. For instance, a translational bioinformatics method was applied to analyze data from formalin-fixed paraffin-embedded tissue microarrays (Sharaf et al., 2011). Correlations were identified among fatty acid–binding protein-1 (FABP-1), pancreatic adenocarcinoma (PaC), and PaC-associated diabetes. These newly identified correlations were suggested as potential useful biomarkers in gastroenterology.

In another example, parsimony phylogenetic analysis has been suggested appropriate for disease hierarchical classification (Abu-Asab et al., 2011). Using this method, the common expressions or mutations are defined in the term of "synapomorphies" to support biomarker discovery. A "synapomorphy" indicates a commonly derived property that may distinguish a certain group from other organisms. The data mining strategies such as parsimonious clustering can be applied for constructing the "omics" profiles for further systems-based studies.

More importantly, the modeling of interacting components and complex interrelationships would enable companion diagnostics and novel treatment strategies (Caberlotto and Lauria, 2015). Companion diagnostic methods may get information on the safety and effectiveness of treatment agents including drugs by applying medical devices.

In the cases of musculoskeletal disorders such as rheumatoid arthritis, the complex and multifactorial features may complicate patient-specific therapies (Gibson et al., 2015). These obstacles have led to low effectiveness and failed therapeutic outcomes. Systems and dynamical biomarkers based on the integration of proteomics may help to tackle the difficulties by effective examinations of joint impairments, autoimmune status, and disease development. Strategies of "companion diagnostics" may be helpful for promoting such biomarker discovery, drug selections, and treatment results. Such assessments can be used as powerful decision support approaches for individualized interventions.

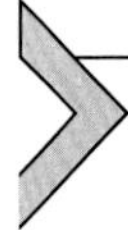

5.3 FINDING ROBUST BIOMARKERS FOR SYSTEMS AND DYNAMICAL MEDICINE

Translational bioinformatics and systems biology methods such as supervised machine learning and HTP data mining can be especially useful for the discovery of robust biomarkers in complex diseases such as the host–pathogen interaction networks (Dix et al., 2016). By viewing the complex systems as a whole, the identification of putative biomarkers would be possible for multidimensional diagnosis and therapeutic decision support (see Chapters 2 and 4). The network modeling would help to identify putative drug targets.

More accurate and robust biomarkers may be useful for addressing multiple factors in systems and dynamical medicine. These factors include genomic and epigenetic variations, functional changes, the structure–function and genotype–phenotype correlations across spatial levels, the etiologic diversity across temporal scales, and environmental effects (Filiou and Turck, 2011; also see Fig. 5.1). Specifically, the alterations at the cellular levels may be assessed quantitatively via high content phenotypic tests including the profiling of altered transcriptome or proteome in the whole cell (Dunn et al., 2010). The establishment of the profiles of molecules, peptides, polynucleotides, and pathways may help to identify the perturbing factors as the systems-based biomarkers.

Biomarkers based on predictive modeling may represent both spatial status and temporal phases of illnesses (see Chapter 7). For example, biomarkers highlighting the temporal evolving processes may promote the understanding of the occurrence and development of the clinical manifestations in Alzheimer's disease (Jack et al., 2013). According to the dynamical models, the temporal factors rather than the symptom severities are more suitable for indicating the disease development among different patients.

The HTP technologies have been proven useful for the dynamical examinations and across-scale assessments about the systems-based genotype–phenotype correlations with time-series tests (Chen et al., 2012). As the deteriorations in complex diseases usually happen suddenly at a tipping point featuring an imminent bifurcation, early warning signs need to be detected at this pivotal point. The genome-wide expression profiling based on such approaches may reveal disease onset and progression for defining the next-generation biomarkers (Wang et al., 2013). The strategies such as the "virtual patient" models may emphasize the comprehensive interactions with the virtualization of potential drug targets for personalized medicine (Wierling et al., 2015).

Specifically, the dynamical network biomarkers (DNBs) can be identified from the HTP gene expression data (Chen et al., 2012). The tissue-specific molecules in the DNBs can be assessed from the normal status to the disease condition within the whole organism (Li et al., 2014). Such strategies would improve early diagnosis and more precise prognosis. Dynamical methods such as the modeling of clonal evolution of biomarkers may provide immediate detection of cancers for better clinical management (Li et al., 2011). A series of robust biomarkers would promote the prediction of cancer development with the stratification of patient subgroups for timely and individualized treatment.

Such dynamical approaches can address the deep impacts of the temporal factors on the human organs including both structural and functional interactions (Pantelis et al., 2005; Rapoport et al., 2005; Iris, 2008). At the molecular level, gene expression alterations are associated with the functional evolutions over time such as the aging processes (Pearce et al., 2007; Shames et al., 2007; also see Chapter 12). These spatiotemporal alterations are influential in all of the human systems especially the central nervous system (Shen et al., 2006; Popesco et al., 2007; Iris, 2008).

In summary, robust biomarkers may help to identify such dynamical and complex alterations because the functional status at one time point may be distinct from another time point. For instance, the elucidation of network dynamics may be the key in the studies of prion disease and drug-induced liver injury (Wang et al., 2010). Methods such as in vitro experimental perturbation and systems biology modeling may help to explore the conditions of the complex networks. New and more effective drug targets can be identified for the management of these dynamical conditions to prevent disease development. These strategies may promote the development of systems and dynamical medicine.

REFERENCES

Abu-Asab, M.S., Chaouchi, M., Alesci, S., et al., 2011. Biomarkers in the age of omics: time for a systems biology approach. OMICS 15, 105–112.

Aich, P., Babiuk, L.A., Potter, A.A., Griebel, P., 2009. Biomarkers for prediction of bovine respiratory disease outcome. OMICS 13 (3), 199–209.

Azuaje, F., Zheng, H., Camargo, A., Wang, H., 2011. Systems-based biological concordance and predictive reproducibility of gene set discovery methods in cardiovascular disease. J. Biomed. Inform. 44 (4), 637–647.

Biomarkers Definitions Working Group, 2001. Biomarkers and surrogate endpoints: preferred definitions and conceptual framework. Clin. Pharmacol. Ther. 69, 89–95.

Butz, H., Szabó, P.M., Nofech-Mozes, R., Rotondo, F., Kovacs, K., Mirham, L., Girgis, H., Boles, D., Patocs, A., Yousef, G.M., 2014. Integrative bioinformatics analysis reveals new prognostic biomarkers of clear cell renal cell carcinoma. Clin. Chem. 60, 1314–1326.

Caberlotto, L., Lauria, M., 2015. Systems biology meets -omic technologies: novel approaches to biomarker discovery and companion diagnostic development. Expert Rev. Mol. Diagn. 15, 255–265. http://dx.doi.org/10.1586/14737159.2015.975214.

Chau, C.H., Rixe, O., McLeod, H., Figg, W.D., 2008. Validation of analytic methods for biomarkers used in drug development. Clin. Cancer Res. 14, 5967–5976.

Chen, L., Liu, R., Liu, Z.-P., et al., 2012. Detecting early-warning signals for sudden deterioration of complex diseases by dynamical network biomarkers. Sci. Rep. 2, 342.

Chia, S., Senatore, F., Raffel, O.C., Lee, H., Wackers, F.J.T., Jang, I.-K., 2008. Utility of cardiac biomarkers in predicting infarct size, left ventricular function, and clinical outcome after primary percutaneous coronary intervention for ST-segment elevation myocardial infarction. JACC Cardiovasc. Interv. 1 (4), 415–423.

Day, M., Rutkowski, J.L., Feuerstein, G.Z., 2009. Translational medicine–a paradigm shift in modern drug discovery and development: the role of biomarkers. Adv. Exp. Med. Biol. 655, 1–12.

Deeks, J.J., Altman, D.G., 2004. Diagnostic tests 4: likelihood ratios. BMJ. 329, 168–169.

Dix, A., Vlaic , S., Guthke, R., Linde, J., 2016. Use of systems biology to decipher host-pathogen interaction networks and predict biomarkers. Clin. Microbiol. Infect. 22, 600–606.

Dunn, D.A., Apanovitch, D., Follettie, M., et al., 2010. Taking a systems approach to the identification of novel therapeutic targets and biomarkers. Curr. Pharm. Biotechnol. 11, 721–734.

Eissa, S., Matboli, M., Bekhet, M.M., 2016. Clinical verification of a novel urinary microRNA panel: 133b, -342 and -30 as biomarkers for diabetic nephropathy identified by bioinformatics analysis. Biomed. Pharmacother. 83, 92–99.

Emerson, J.W., Dolled-Filhart, M., Harris, L., Rimm, D.L., Tuck, D.P., 2009. Quantitative assessment of tissue biomarkers and construction of a model to predict outcome in breast cancer using multiple imputation. Cancer Inform. 7, 29–40.

Fall, D.J., Stessman, H., Patel, S.S., Sachs, Z., Van Ness, B.G., Baughn, L.B., Linden, M.A., 2014. Utilization of translational bioinformatics to identify novel biomarkers of bortezomib resistance in multiple myeloma. J. Cancer 5, 720–727.

Filiou, M.D., Turck, C.W., 2011. General overview: biomarkers in neuroscience research. Int. Rev. Neurobiol. 101, 1–17.

Fu, Q., Schoenhoff, F.S., Savage, W.J., Zhang, P., Van Eyk, J.E., 2010. Multiplex assays for biomarker research and clinical application: translational science coming of age. Proteom. Clin. Appl. 4 (3), 271–284.

Gibson, D.S., Bustard, M.J., McGeough, C.M., et al., 2015. Current and future trends in biomarker discovery and development of companion diagnostics for arthritis. Expert Rev. Mol. Diagn. 15, 219–234. http://dx.doi.org/10.1586/14737159.2015.969244.

Hurks, R., Peeters, W., Derksen, W.J.M., Hellings, W.E., Hoefer, I.E., Moll, F.L., de Kleijn, D.P.V., et al., 2009. Biobanks and the search for predictive biomarkers of local and systemic outcome in atherosclerotic disease. Thromb. Haemost. 101 (1), 48–54.

Iris, F., 2008. Biological modeling in the discovery and validation of cognitive dysfunctions biomarkers. In: Turck, C.W. (Ed.), Biomarkers for Psychiatric Disorders. Springer Science+Business Media, New York.

Jack Jr., C.R., Knopman, D.S., Jagust, W.J., et al., 2013. Tracking pathophysiological processes in Alzheimer's disease: an updated hypothetical model of dynamic biomarkers. Lancet Neurol. 12, 207–216.

Knudsen, L.S., Klarlund, M., Skjødt, H., Jensen, T., Ostergaard, M., Jensen, K.E., Hansen, M.S., et al., 2008. Biomarkers of inflammation in patients with unclassified polyarthritis and early rheumatoid arthritis. Relationship to disease activity and radiographic outcome. J. Rheumatol. 35 (7), 1277–1287.

Li, M., Zeng, T., Liu, R., Chen, L., 2014. Detecting tissue-specific early warning signals for complex diseases based on dynamical network biomarkers: study of type 2 diabetes by cross-tissue analysis. Brief. Bioinform 15, 229–243.

Li, X., Blount, P.L., Vaughan, T.L., Reid, B.J., 2011. Application of biomarkers in cancer risk management: evaluation from stochastic clonal evolutionary and dynamic system optimization points of view. PLoS Comput. Biol. 7, e1001087.

Markman, B., Tabernero, J., 2013. Biomarkers for Go/No Go decisions. In: Lenz, H.-J. (Ed.), Biomarkers in Oncology. Prediction and Prognosis, Springer Science+Business Media, New York.

Ozkisacik, E.A., Discigil, B., Boga, M., Gurcun, U., Badak, M.I., Kurtoglu, T., Yenisey, C., et al., 2006. Effects of cyclosporin A on neurological outcome and serum biomarkers in the same setting of spinal cord ischemia model. Ann. Vasc. Surg. 20 (2), 243–249.

Pantelis, C., Yucel, M., et al., 2005. Structural brain imaging evidence for multiple pathological processes at different stages of brain development in schizophrenia. Schizophr. Bull. 31, 672–696.

Pearce, D.J., Anjos-Afonso, F., et al., 2007. Age-dependent increase in side population distribution within hematopoiesis: implications for our understanding of the mechanism of aging. Stem Cell 25, 828–835.

Popesco, M.C., Lin, S., et al., 2007. Serial analysis of gene expression profiles of adult and aged mouse cerebellum. Neurobiol. Aging 29, 774–788.

Puchades-Carrasco, L., Palomino-Schätzlein, M., Pérez-Rambla, C., Pineda-Lucena, A., 2016. Bioinformatics tools for the analysis of NMR metabolomics studies focused on the identification of clinically relevant biomarkers. Brief. Bioinform. 17, 541–552.

Rapoport, J.L., Addington, A.M., et al., 2005. The neurodevelopmental model of schizophrenia: update 2005. Mol. Psychiatr. 10, 434–449.

Shames, D.S., Minna, J.D., et al., 2007. DNA methylation in health, disease, and cancer. Curr. Mol. Med. 7, 85–102.

Sharaf, R.N., Butte, A.J., Montgomery, K.D., Pai, R., Dudley, J.T., Pasricha, P.J., 2011. Computational prediction and experimental validation associating FABP-1 and pancreatic adenocarcinoma with diabetes. BMC Gastroenterol. 11, 5.

Shen, S., Liu, A., et al., 2006. Epigenetic memory loss in aging oligodendrocytes in the corpus callosum. Neurobiol. Aging 29, 452–463.

Vasan, R.S., 2006. Biomarkers of cardiovascular disease: molecular basis and practical considerations. Circulation 113, 2335–2362.

Vilar, E., Tabernero, J., 2013. Biomarker discovery strategies: DNA, RNA, and protein. In: Lenz, H.-J. (Ed.), Biomarkers in Oncology: Prediction and Prognosis. Springer Science+Business Media, New York.

Wang, K., Lee, I., Carlson, G., Hood, L., Galas, D., , 2010. Systems biology and the discovery of diagnostic biomarkers. Dis. Markers 28, 199–207.

Wang, I.M., Stone, D.J., Nickle, D., et al., 2013. Systems biology approach for new target and biomarker identification. Curr. Top Microbiol. Immunol. 363, 169–199. http://dx.doi.org/10.1007/82_2012_252.

Wehling, M., 2008. Translational medicine: science or wishful thinking? J. Transl. Med. 6, 31.

Wierling, C., Kessler, T., Ogilvie, L.A., et al., 2015. Network and systems biology: essential steps in virtualising drug discovery and development. Drug Discov. Today Technol. 15, 33–40. http://dx.doi.org/10.1016/j.ddtec.2015.07.002.

Yan, Q., 2011. Toward the integration of personalized and systems medicine: challenges, opportunities and approaches. Pers. Med. 8, 1–4.

CHAPTER SIX

Biomarkers From Systems Biology and "Omics" Studies: Applications and Examples

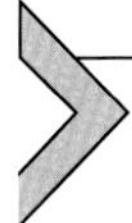

6.1 PROTEOMIC AND METABOLOMIC PATHWAYS AND BIOMARKERS

Proteomics and metabolomics studies are essential in systems biology (see Chapter 3). The analyses of data from these studies may promote the accuracy, sensitivity, and throughput for biomarker identification because the proteome represents the functional actors in a cell. The integrative analyses at the proteome level may contribute to the systemic profiling to support the systems-based discovery of biomarkers (Pitteri and Hanash, 2010).

For instance, gender-specific cytokine pathways have been identified as the potential biomarkers and drug targets for individualized cancer therapy (Berghella et al., 2016; also see Chapter 11). Such biomarkers may be applied for stratifying the patients for adjuvant therapies during the early stages of cancer as well as those subgroups during the advanced pathological stages. The interrelationships among the cytokine pathways associated with the disease progression have been proposed for personalized interventions for cancers.

High-throughput (HTP) technologies such as protein microarrays can be useful for proteomic studies to support translational medicine with the integration of scientific studies and bedside practice. Different kinds of protein microarrays can be utilized for the examinations of smaller peptides, antibodies, and the whole proteomes (Tu et al., 2014). Such integrative approaches may enable the parallel assessments for the large-scale evaluations of the complex protein–protein–drug interactions and antigen–antibody interactions. As an example, such strategies were proven useful for the finding of systems-grounded biomarkers based on the profiling of live cell surface glycan (Tu et al., 2014).

Different activations of signaling pathways associated with various cell actions have been suggested as mechanism-based robust biomarkers for the

Translational Bioinformatics and Systems Biology Methods for Personalized Medicine
ISBN 978-0-12-804328-8
http://dx.doi.org/10.1016/B978-0-12-804328-8.00006-1

establishment of predictive frameworks about drug sensitivities (Amadoz et al., 2015). In the case of cancer, various cell lines can be utilized as the model system for identifying proteomic biomarkers. For instance, the profiling of plasma proteomes has been suggested to improve the detection of the protein alterations between normal and breast cancer tissues (Zhang and Chen, 2010). Such comparisons and tracking of alterations at the proteomic levels may be especially helpful for identifying systems-based robust biomarkers (see Chapters 5 and 11).

Various translational bioinformatics methods can be applied for the identification and validation of biomarkers, including the statistics and permutation approaches, pathway and functional modeling, and cross-validation of multiple analyses (see Chapter 5). For instance, data mining of the proteomics data set from the plasma samples of breast cancer patients and healthy controls successfully identified 254 abnormally expressed proteins (Zhang and Chen, 2010). The combination of bioinformatics and experimental tests showed 25 proteins and their associated pathways as the potential biomarkers, such as those related to coagulation cascades and actin cytoskeleton. Using gene ontology annotations, correlations have also been confirmed with metabolic mechanisms, proteolysis, and acute inflammation. Moreover, the cross-validation indicated strong connections in the pathway–protein matrix. In the following section, some examples of the pathways as potential biomarkers will be discussed.

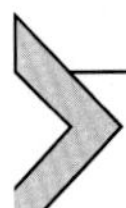

6.2 PATHWAYS AS POTENTIAL BIOMARKERS: EXAMPLES

Table 6.1 shows some examples of the potential pathway-based biomarkers for various diseases. A more complete and updated list can be found at the site of Biomarkers and Systems Medicine (BSM, 2016). For example, the nuclear factor kappa B (NF-κB) signaling pathway has critical roles in many diseases such as age-associated diseases and AIDS-associated Burkitt lymphoma (Balistreri et al., 2013; Ramos et al., 2012). As the potential biomarker, NF-κB signaling pathway activators have been suggested as the therapeutic targets for aging and age-associated diseases.

The phosphoinositide 3-kinase (PI3K) pathway has been related to head and neck squamous cell carcinoma (HNSCC), breast cancer, prostate cancer, and pediatric ependymoma (Lui et al., 2013; Kremer et al., 2006; Rogers et al., 2013) (see Table 6.1). The PI3K/AKT/mTOR pathways may have significant roles for breast cancer targets and biomarkers (Paplomata and O'Regan, 2014).

Table 6.1 Examples of Potential Pathway Biomarkers in Various Diseases

Potential Biomarkers	Associated Diseases	References
EGFR pathways	CRC	Pierobon et al. (2009)
	Glial tumors	Comincini et al. (2009)
	Localized squamous laryngeal carcinoma	Dionysopoulos et al. (2013)
	Pancreatic cancer	Boeck et al. (2013)
	Skin rash in metastatic breast cancer treated with erlotinib	Tan et al. (2008)
	Synovial sarcoma	Teng et al. (2011)
IGF signaling pathways	Human sarcomas treatment	Lin et al. (2013)
	Invasive breast cancer	Lawlor et al. (2009)
	Metastatic breast cancer treated with cetuximab	Huang et al. (2012)
	NSCLC	Shersher et al. (2011)
mTOR signaling pathways	ccRCC; nonmetastatic kidney cancer	Darwish et al. (2013)
	HNSCC	Clark et al. (2010)
	Renal cell carcinoma, after radical nephrectomy	Nishikawa et al. (2014)
NF-κB signaling pathways and activators	Aging and age-associated diseases	Balistreri et al. (2013)
	AIDS-associated BL	Ramos et al. (2012)
	Diffuse large B-cell lymphoma	Thompson et al. (2011)
PI3K/AKT/mTOR pathways	HNSCC	Lui et al. (2013)
	Pediatric ependymoma	Rogers et al. (2013)
	Prostate cancer	Kremer et al. (2006)
TNF pathways	Depression, SSRI treatment responses	Powell et al. (2013)
WNT/β-catenin pathways	Advanced epithelial ovarian cancer	Bodnar et al. (2014)
	Cellular replicative senescence	Binet et al. (2009)
	CRC	Ting et al. (2013)
	Epithelial ovarian cancer	Dai et al. (2011)
	Glioblastoma (brain tumor)	Zhu et al. (2013)
	IBD-associated colorectal carcinogenesis	Claessen et al. (2010)
	Juvenile nasopharyngeal angiofibroma	Ponti et al. (2008)

BL, Burkitt lymphoma; *ccRCC*, clear cell renal cell carcinoma; *CRC*, colorectal cancer; *EGFR*, epidermal growth factor receptor; *HNSCC*, head and neck squamous cell carcinoma; *IBD*, inflammatory bowel disease; *IGF*, insulin-like growth factor; *mTOR*, mammalian target of rapamycin; *NSCLC*, non–small cell lung cancer; *PI3K*, phosphoinositide 3-kinase; *SSRI*, selective serotonin reuptake inhibitor; *TNF*, tumor necrosis factor.

The mammalian target of rapamycin (mTOR) pathways have been associated with clear cell renal cell carcinoma (RCC) and HNSCC (Darwish et al., 2013; Clark et al., 2010). They have been suggested as the potential biomarkers for the prediction of disease recurrence after radical nephrectomy for nonmetastatic RCC, as well as for diagnosis and therapeutic responses of lung cancer (Darwish et al., 2013; Nishikawa et al., 2014) (see Table 6.1).

The insulin-like growth factor (IGF) signaling pathway may be the potential biomarkers for human sarcomas treatment (Lin et al., 2013). The IGF signaling and the plasminogen activating system may be the potential prognostic and predictive biomarkers for invasive breast cancer (Lawlor et al., 2009). The HTP proteomics profiling of secretomes may be helpful for such identification. This pathway may also be used for the detection of tumor progression and analysis of patient outcomes in non–small cell lung cancer (NSCLC) (Shersher et al., 2011) (see Table 6.1).

Genes associated with the epidermal growth factor receptor (EGFR) pathway have been suggested as the potential biomarkers for the differential diagnosis of high-grade gliomas and the prognosis of NSCLC and synovial sarcoma (Comincini et al., 2009; Teng et al., 2011). The EGFR pathway has also been proposed to be the potential prognostic or predictive biomarkers for pancreatic cancer (Boeck et al., 2013) (see Table 6.1).

In addition, members of the EGFR and COX2 signaling pathways may be the potential prognostic biomarkers for the prediction of occult metastasis in colorectal cancer (Pierobon et al., 2009). Multiple pathways including EGFR, cyclin D1, and Akt/mTOR pathways may be the potential prognostic markers for localized squamous laryngeal carcinoma (Dionysopoulos et al., 2013) (see Table 6.1).

Genes associated with the WNT pathway, angiogenetic and hormonal factors have been associated with juvenile nasopharyngeal angiofibroma (Ponti et al., 2008). The DNA methylation at promoter CpG islands associated with the WNT pathway has been suggested as the potential predictive biomarkers of patient progression-free survival in epithelial ovarian cancer (EOC) (Dai et al., 2011).

The WNT/β-catenin pathway and E-cadherin have also been related to advanced epithelial ovarian cancer (AEOC) and inflammatory bowel disease (IBD)–associated colorectal carcinogenesis as a potential prognostic biomarker (Bodnar et al., 2014; Claessen et al., 2010). The moesin and CD44 have important roles in the activation of the WNT/β-catenin pathway. They have been suggested as the potential progression biomarkers for glioblastoma (brain tumor) (Zhu et al., 2013) (see Table 6.1). WNT16B may regulate p53 and the PI3K/AKT pathway activity and has been suggested as a biomarker for the cellular replicative senescence (Binet et al., 2009) (see Table 6.1).

The tumor necrosis factor (TNF) and targets in the inflammatory cytokine pathway have been associated with depression and the responses to the selective serotonin reuptake inhibitor (SSRI) antidepressant escitalopram. They have been suggested as the potential predictive biomarkers for the treatment responses (Powell et al., 2013) (see Table 6.1).

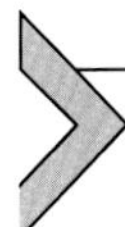

6.3 POTENTIAL MICRORNA BIOMARKERS AND EXAMPLES

In addition to proteomic and pathway biomarkers, extracellular microRNAs (miRNAs) have been suggested as promising biomarkers. For instance, the application of miRNAs as the noninvasive detector and biomarkers for disease progression and treatment responses was considered useful for Duchenne muscular dystrophy (Roberts et al., 2013). The profiling of serum miRNAs and the extracellular miRNA markers may represent the dose-responsive restoration following dystrophin rescue.

Furthermore, the dynamical expression patterns were detected in extracellular dystrophy-associated miRNAs (dystromiRs) that may characterize the progression of muscle pathology. The serum dystromiR levels were related to experimentally induced skeletal muscle injury, the expression of the myogenic miR-206, as well as the myogenic transcription factor myogenin (Roberts et al., 2013). These examinations have demonstrated that extracellular miRNAs may be used as dynamical biomarkers for the regenerative conditions of the musculature to support the exploration of pathophysiological mechanisms in skeletal muscles.

Table 6.2 lists some examples of miRNAs as the potential biomarkers for various diseases. A more complete and updated list can be found at the site of Biomarkers and Systems Medicine (BSM, 2016). For instance, the upregulation of miR-494, miR-1973, and miR-21 may be the potential disease response biomarkers for classical Hodgkin lymphoma (Jones et al., 2014). Keratin-18 and miRNA-122 complement alanine aminotransferase have been suggested as the predictive and safety biomarkers for drug-induced liver injury (Thulin et al., 2014) (see Table 6.2).

In addition, miR-31, miR-206, miR-424, and miR-146a may be the potential diagnostic biomarkers for IBD (Lin et al., 2014). MiR-21, miR-106b, miR-17, miR-18a, and miR-20a may be the potential diagnostic and/or prognostic biomarkers for gastric cancer (Wang et al., 2013). Upregulated members of the miR-200 family and the downregulation of miR-100 may be used as the potential biomarkers for the early diagnosis of EOC (Chen et al., 2013) (see Table 6.2).

As shown in these examples, multiple miRNAs and associated networks are usually found useful as potential biomarkers. For instance, the serum miRNA profiles including miR-17, miR-18a, and miR-20a as well as the key regulatory genes of cell proliferation, apoptosis, and regulatory networks may be the possible diagnostic biomarkers for retinoblastoma (Beta et al., 2013) (see Table 6.2).

Urinary miRNAs, miR-618 and miR-650, have been considered promising as the screening biomarkers for the early detection of hepatitis C virus (HCV)–associated hepatocellular carcinoma (Abdalla and Haj-Ahmad, 2012). Higher levels of the serum-based miR-1254 and miR-574-5p may be used as the potential biomarkers for the early diagnosis of NSCLC (Foss et al., 2011) (see Table 6.2).

The combinations of different molecules may be especially helpful. For example, a panel of 10 protein-coding genes and two miRNA genes including CCNG2, NOTCH3, miR-519d, and miR-647 were suggested as the potential prognostic biomarkers for the biochemical recurrence among prostate cancer patients after radical prostatectomy. They have been found useful for differentiating patients with and without biochemical recurrence (Long et al., 2011) (see Table 6.2).

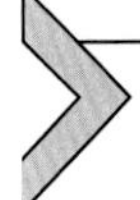

6.4 DYNAMICAL CIRCADIAN BIOMARKERS AND CHRONOTHERAPY

As discussed previously, biomarkers may have broad and significant impacts on the translation of the pathological studies into clinical practice. The discovery of specific dynamical biomarkers may not only enable the classification of patients into different clinical and therapeutic subgroups but also help with precise diagnosis, prevention, and therapies during different stages of diseases (see Chapters 1, 2, and 9–12). To meet these objectives, it is essential to find biomarkers that represent disease progress and evolvement timely to be used for drug target discovery and the prediction of disease states and treatment responses.

Table 6.3 shows some examples of potential biomarkers associated with temporal changes. A more complete and updated list can be found at the site of Biomarkers and Systems Medicine (BSM, 2016). For example, circadian patterns have been identified in the salivary levels of melatonin and cortisol. The daily profiles of melatonin and cortisol have been related

Table 6.2 Examples of Potential miRNA Biomarkers in Various Diseases

Associated Diseases	Potential Biomarkers	References
Blood stasis syndrome in unstable angina	MiR-146b-5p, miR-199a-5p, CALR, and TP53	Wang and Yu (2013)
Classical Hodgkin lymphoma	MiR-494, miR-1973 and miR-21	Jones et al. (2014)
Chronic lymphocytic leukemia	Circulating miR-20a	Moussay et al. (2011)
Drug-induced liver injury	M65 and microRNA-122	Thulin et al. (2014)
Eosinophilic esophagitis	Esophageal miRNAs miR-146a, miR-146b, and miR-223	Lu et al. (2012)
Epithelial ovarian cancer	MiR-200a, miR-200b, miR-200c, and miR-141; miR-100	Chen et al. (2013)
Gastric cancer	MiR-21, miR-106b, miR-17, miR-18a and miR-20a	Wang et al. (2013)
Hepatitis C virus–associated HCC	MiR-618, miR-650	Abdalla and Haj-Ahmad (2012)
HCCs, surgically resected	MicroRNA-221, microRNA-222, microRNA-21, and microRNA-155	Yoon et al. (2011)
Inflammatory bowel disease (ulcerative colitis, Crohn's disease)	MiR-31, miR-206, miR-424, and miR-146a	Lin et al. (2014)
Diffuse large B-cell lymphoma	MiR-24	Culpin et al. (2013)
Lung cancer, malignant solitary pulmonary nodules	MiR-21 and miR-210, miR-486-5p	Shen et al. (2011)
Multiple sclerosis	Let-7g and miR-150	Martinelli-Boneschi et al. (2012)
NSCLC	Serum-based miR-1254 and miR-574-5p	Foss et al. (2011)
Ovarian cancer	Serum miR-132, miR-26a, let-7b, and miR-145	Chung et al. (2013)
Primary breast cancer	MiR-92a and miR-21	Si et al. (2013)
Prostate cancer	10 protein-coding genes and 2 miRNA genes (RAD23B, FBP1, TNFRSF1A, CCNG2, NOTCH3, ETV1, BID, SIM2, LETMD1, ANXA1, miR-519d, and miR-647)	Long et al. (2011)
Retinoblastoma	MiR-17, miR-18a, miR-20a, and regulatory genes of cell proliferation, apoptosis	Beta et al. (2013)

HCC, hepatocellular carcinoma; *NSCLC*, Non–small cell lung cancer.

to metabolic syndrome components. The altered patterns may be used as the potential biomarkers for the metabolic and homeostatic disturbances in blood pressure, glucose, and plasma lipids regulations (Corbalán-Tutau et al., 2014).

The diurnal serum levels of cortisol and inflammatory markers such as soluble TNF-β and soluble vascular adhesion molecule-1 (sVCAM-1) have been associated with multiple sclerosis (MS). The circadian patterns of the relevant inflammatory serum parameters have been suggested as the potential biomarkers for patients with MS (Wipfler et al., 2013) (see Table 6.3).

In addition, abnormal quantitative and qualitative circadian patterns and circadian genes have been detected using various methods such as polysomnography and blood melatonin monitoring. These circadian patterns may indicate the trait and altered physiological functions. They may be applied as the potential dynamical biomarkers for bipolar disorders (Milhiet et al., 2011) (see Table 6.3).

Studies in chronobiology have discovered that genes involved in circadian rhythms are the critical components of the pathways related to cell proliferation, cell cycles, and apoptosis (Zhu et al., 2005). The alterations in circadian genes may lead to tumorigenesis and cancer development, such as neuronal PAS domain protein 2 (NPAS2) (Yi et al., 2010). A missense polymorphism in NPAS2 (Ala394Thr) may be involved in the higher risks of tumors such as breast cancer (see Table 6.3). These factors have been suggested as possible prognostic biomarkers. In another example, the structural variations in the circadian gene Period3 (PER3) have been considered as the potential biomarkers for breast cancer among young women (Zhu et al., 2005; also see Table 6.3).

The skin surface temperature rhythms have been proposed as the promising circadian biomarkers for personalized chronotherapeutics among cancer patients (Scully et al., 2011). The timing of anticancer medication administration based on the circadian rhythms has been proven to promote the treatment tolerability up to fivefold and double the treatment efficacy (see Table 6.3). Multiple skin temperature locations may be necessary to decide the accurate circadian stage and the individualized chronotherapeutic plans.

Table 6.3 Examples of Potential Dynamical Biomarkers in Various Diseases

Associated Diseases/ Conditions	Potential Biomarkers	References
Bipolar disorders	Circadian disturbances and abnormal blood melatonin levels; circadian genes	Milhiet et al. (2011)
Breast cancer	NPAS2 and other circadian genes; a polymorphism in NPAS2 (Ala394Thr)	Yi et al. (2010)
Breast cancer in young women	PER3 structural variations	Zhu et al. (2005)
Cancer; personalized chronotherapeutics	Skin surface temperature rhythms	Scully et al. (2011)
Metabolic syndrome including disturbances in blood pressure, glucose, and lipid regulations	Salivary melatonin and cortisol circadian patterns	Corbalán-Tutau et al. (2014)
Multiple sclerosis	Diurnal patterns in cortisol and inflammatory markers including sTNF-β, sTNF-R1, sTNF-2, sVCAM-1, sICAM-1	Wipfler et al. (2013)

REFERENCES

Abdalla, M.A., Haj-Ahmad, Y., 2012. Promising candidate urinary microRNA biomarkers for the early detection of hepatocellular carcinoma among high-risk hepatitis C virus Egyptian patients. J. Cancer 3, 19–31.

Amadoz, A., Sebastian-Leon, P., Vidal, E., Salavert, F., Dopazo, J., 2015. Using activation status of signaling pathways as mechanism-based biomarkers to predict drug sensitivity. Sci. Rep. 5, 18494.

Balistreri, C.R., Candore, G., Accardi, G., Colonna-Romano, G., Lio, D., 2013. NF-κB pathway activators as potential ageing biomarkers: targets for new therapeutic strategies. Immun. Ageing 10, 24.

Berghella, A.M., Contasta, I., Lattanzio, R., Di Gregorio, G., Campitelli, I., Marino, S., Liberatore, L.L., Navarra, L., Caterino, G., Mongelli, A., et al., 2016. The role of gender-specific cytokine pathways as drug targets and gender-specific biomarkers in personalized cancer therapy. Curr. Drug Targets.

Beta, M., Venkatesan, N., Vasudevan, M., Vetrivel, U., Khetan, V., Krishnakumar, S., 2013. Identification and insilico analysis of retinoblastoma serum microRNA profile and gene targets towards prediction of novel serum biomarkers. Bioinform. Biol. Insights 7, 21–34.

Binet, R., Ythier, D., Robles, A.I., Collado, M., Larrieu, D., Fonti, C., Brambilla, E., Brambilla, C., Serrano, M., Harris, C.C., et al., 2009. WNT16B is a new marker of cellular senescence that regulates p53 activity and the phosphoinositide 3-kinase/AKT pathway. Cancer Res. 69, 9183–9191.

Bodnar, L., Stanczak, A., Cierniak, S., Smoter, M., Cichowicz, M., Kozlowski, W., Szczylik, C., Wieczorek, M., Lamparska-Przybysz, M., 2014. Wnt/β-catenin pathway as a potential prognostic and predictive marker in patients with advanced ovarian cancer. J. Ovarian Res. 7, 16.

Boeck, S., Jung, A., Laubender, R.P., Neumann, J., Egg, R., Goritschan, C., Vehling-Kaiser, U., Winkelmann, C., Fischer von Weikersthal, L., Clemens, M.R., et al., 2013. EGFR pathway biomarkers in erlotinib-treated patients with advanced pancreatic cancer: translational results from the randomised, crossover phase 3 trial AIO-PK0104. Br. J. Cancer 108, 469–476.

BSM, 2016. Biomarkers and Systems Medicine. http://pharmtao.com/health/category/systems-medicine/biomarkers-systems-medicine.

Chen, Y., Zhang, L., Hao, Q., 2013. Candidate microRNA biomarkers in human epithelial ovarian cancer: systematic review profiling studies and experimental validation. Cancer Cell Int. 13, 86.

Chung, Y.-W., Bae, H.-S., Song, J.-Y., Lee, J.K., Lee, N.W., Kim, T., Lee, K., 2013. Detection of microRNA as novel biomarkers of epithelial ovarian cancer from the serum of ovarian cancer patients. Int. J. Gynecol. Cancer 23, 673–679.

Claessen, M.M.H., Schipper, M.E.I., Oldenburg, B., Siersema, P.D., Offerhaus, G.J.A., Vleggaar, F.P., 2010. WNT-pathway activation in IBD-associated colorectal carcinogenesis: potential biomarkers for colonic surveillance. Cell. Oncol. 32, 303–310.

Clark, C., Shah, S., Herman-Ferdinandez, L., Ekshyyan, O., Abreo, F., Rong, X., McLarty, J., Lurie, A., Milligan, E.J., Nathan, C.-A.O., 2010. Teasing out the best molecular marker in the AKT/mTOR pathway in head and neck squamous cell cancer patients. Laryngoscope 120, 1159–1165.

Comincini, S., Paolillo, M., Barbieri, G., Palumbo, S., Sbalchiero, E., Azzalin, A., Russo, M.A., Schinelli, S., 2009. Gene expression analysis of an EGFR indirectly related pathway identified PTEN and MMP9 as reliable diagnostic markers for human glial tumor specimens. J. Biomed. Biotechnol. 2009, 924565.

Corbalán-Tutau, D., Madrid, J.A., Nicolás, F., Garaulet, M., 2014. Daily profile in two circadian markers "melatonin and cortisol" and associations with metabolic syndrome components. Physiol. Behav. 123, 231–235.

Culpin, R.E., Sieniawski, M., Proctor, S.J., Menon, G., Mainou-Fowler, T., 2013. MicroRNAs are suitable for assessment as biomarkers from formalin-fixed paraffin-embedded tissue, and miR-24 represents an appropriate reference microRNA for diffuse large B-cell lymphoma studies. J. Clin. Pathol. 66, 249–252.

Dai, W., Teodoridis, J.M., Zeller, C., Graham, J., Hersey, J., Flanagan, J.M., Stronach, E., Millan, D.W., Siddiqui, N., Paul, J., et al., 2011. Systematic CpG islands methylation profiling of genes in the wnt pathway in epithelial ovarian cancer identifies biomarkers of progression-free survival. Clin. Cancer Res. 17, 4052–4062.

Darwish, O.M., Kapur, P., Youssef, R.F., Bagrodia, A., Belsante, M., Alhalabi, F., Sagalowsky, A.I., Lotan, Y., Margulis, V., 2013. Cumulative number of altered biomarkers in mammalian target of rapamycin pathway is an independent predictor of outcome in patients with clear cell renal cell carcinoma. Urology 81, 581–586.

Dionysopoulos, D., Pavlakis, K., Kotoula, V., Fountzilas, E., Markou, K., Karasmanis, I., Angouridakis, N., Nikolaou, A., Kalogeras, K.T., Fountzilas, G., 2013. Cyclin D1, EGFR, and Akt/mTOR pathway. Potential prognostic markers in localized laryngeal squamous cell carcinoma. Strahlenther. Onkol. 189, 202–214.

Foss, K.M., Sima, C., Ugolini, D., Neri, M., Allen, K.E., Weiss, G.J., 2011. miR-1254 and miR-574-5p: serum-based microRNA biomarkers for early-stage non-small cell lung cancer. J. Thorac. Oncol. 6, 482–488.

Huang, F., Xu, L.-A., Khambata-Ford, S., 2012. Correlation between gene expression of IGF-1R pathway markers and cetuximab benefit in metastatic colorectal cancer. Clin. Cancer Res. 18, 1156–1166.

Jones, K., Nourse, J.P., Keane, C., Bhatnagar, A., Gandhi, M.K., 2014. Plasma microRNA are disease response biomarkers in classical Hodgkin lymphoma. Clin. Cancer Res. 20, 253–264.

Kremer, C.L., Klein, R.R., Mendelson, J., Browne, W., Samadzedeh, L.K., Vanpatten, K., Highstrom, L., Pestano, G.A., Nagle, R.B., 2006. Expression of mTOR signaling pathway markers in prostate cancer progression. The Prostate 66, 1203–1212.
Lawlor, K., Nazarian, A., Lacomis, L., Tempst, P., Villanueva, J., 2009. Pathway-based biomarker search by high-throughput proteomics profiling of secretomes. J. Proteome Res. 8, 1489–1503.
Lin, F., Shen, Z., Xu, X., Hu, B.-B., Meerani, S., Tang, L.-N., Zheng, S.-E., Sun, Y.-J., Min, D.-L., Yao, Y., 2013. Evaluation of the expression and role of IGF pathway biomarkers in human sarcomas. Int. J. Immunopathol. Pharmacol. 26, 169–177.
Lin, J., Welker, N.C., Zhao, Z., Li, Y., Zhang, J., Reuss, S.A., Zhang, X., Lee, H., Liu, Y., Bronner, M.P., 2014. Novel specific microRNA biomarkers in idiopathic inflammatory bowel disease unrelated to disease activity. Mod. Pathol. 27, 602–608.
Long, Q., Johnson, B.A., Osunkoya, A.O., Lai, Y.-H., Zhou, W., Abramovitz, M., Xia, M., Bouzyk, M.B., Nam, R.K., Sugar, L., et al., 2011. Protein-coding and microRNA biomarkers of recurrence of prostate cancer following radical prostatectomy. Am. J. Pathol. 179, 46–54.
Lu, T.X., Sherrill, J.D., Wen, T., Plassard, A.J., Besse, J.A., Abonia, J.P., Franciosi, J.P., Putnam, P.E., Eby, M., Martin, L.J., et al., 2012. MicroRNA signature in patients with eosinophilic esophagitis, reversibility with glucocorticoids, and assessment as disease biomarkers. J. Allergy Clin. Immunol. 129, 1064–1075 e9.
Lui, V.W.Y., Hedberg, M.L., Li, H., Vangara, B.S., Pendleton, K., Zeng, Y., Lu, Y., Zhang, Q., Du, Y., Gilbert, B.R., et al., 2013. Frequent mutation of the PI3K pathway in head and neck cancer defines predictive biomarkers. Cancer Discov. 3, 761–769.
Martinelli-Boneschi, F., Fenoglio, C., Brambilla, P., Sorosina, M., Giacalone, G., Esposito, F., Serpente, M., Cantoni, C., Ridolfi, E., Rodegher, M., et al., 2012. MicroRNA and mRNA expression profile screening in multiple sclerosis patients to unravel novel pathogenic steps and identify potential biomarkers. Neurosci. Lett. 508, 4–8.
Milhiet, V., Etain, B., Boudebesse, C., Bellivier, F., 2011. Circadian biomarkers, circadian genes and bipolar disorders. J. Physiol. Paris 105, 183–189.
Moussay, E., Wang, K., Cho, J.-H., van Moer, K., Pierson, S., Paggetti, J., Nazarov, P.V., Palissot, V., Hood, L.E., Berchem, G., et al., 2011. MicroRNA as biomarkers and regulators in B-cell chronic lymphocytic leukemia. Proc. Natl. Acad. Sci. U. S. A. 108, 6573–6578.
Nishikawa, M., Miyake, H., Harada, K., Fujisawa, M., 2014. Expression of molecular markers associated with the mammalian target of rapamycin pathway in nonmetastatic renal cell carcinoma: effect on prognostic outcomes following radical nephrectomy. Urol. Oncol. 32, 49 e15–e21.
Paplomata, E., O'Regan, R., 2014. The PI3K/AKT/mTOR pathway in breast cancer: targets, trials and biomarkers. Ther. Adv. Med. Oncol. 6, 154–166.
Pierobon, M., Calvert, V., Belluco, C., Garaci, E., Deng, J., Lise, M., Nitti, D., Mammano, E., De Marchi, F., Liotta, L., et al., 2009. Multiplexed cell signaling analysis of metastatic and nonmetastatic colorectal cancer reveals COX2-EGFR signaling activation as a potential prognostic pathway biomarker. Clin. Colorectal Cancer 8, 110–117.
Pitteri, S., Hanash, S., 2010. A systems approach to the proteomic identification of novel cancer biomarkers. Dis. Markers 28, 233–239.
Ponti, G., Losi, L., Pellacani, G., Rossi, G.B., Presutti, L., Mattioli, F., Villari, D., Wannesson, L., Alicandri Ciufelli, M., Izzo, P., et al., 2008. Wnt pathway, angiogenetic and hormonal markers in sporadic and familial adenomatous polyposis-associated juvenile nasopharyngeal angiofibromas (JNA). Appl. Immunohistochem. Mol. Morphol. 16, 173–178.
Powell, T.R., Schalkwyk, L.C., Heffernan, A.L., Breen, G., Lawrence, T., Price, T., Farmer, A.E., Aitchison, K.J., Craig, I.W., Danese, A., et al., 2013. Tumor necrosis factor and its targets in the inflammatory cytokine pathway are identified as putative transcriptomic biomarkers for escitalopram response. Eur. Neuropsychopharmacol. 23, 1105–1114.

Ramos, J.-C., Sin, S.-H., Staudt, M.R., Roy, D., Vahrson, W., Dezube, B.J., Harrington, W., Dittmer, D.P., 2012. Nuclear factor kappa B pathway associated biomarkers in AIDS defining malignancies. Int. J. Cancer 130, 2728–2733.

Roberts, T.C., Godfrey, C., McClorey, G., Vader, P., Briggs, D., Gardiner, C., Aoki, Y., Sargent, I., Morgan, J.E., Wood, M.J., 2013. Extracellular microRNAs are dynamic non-vesicular biomarkers of muscle turnover. Nucleic Acids Res. 41, 9500–9513.

Rogers, H.A., Mayne, C., Chapman, R.J., Kilday, J.-P., Coyle, B., Grundy, R.G., 2013. PI3K pathway activation provides a novel therapeutic target for pediatric ependymoma and is an independent marker of progression-free survival. Clin. Cancer Res. 19, 6450–6460.

Scully, C.G., Karaboué, A., Liu, W.-M., Meyer, J., Innominato, P.F., Chon, K.H., Gorbach, A.M., Lévi, F., 2011. Skin surface temperature rhythms as potential circadian biomarkers for personalized chronotherapeutics in cancer patients. Interface Focus 1, 48–60.

Shen, J., Liu, Z., Todd, N.W., Zhang, H., Liao, J., Yu, L., Guarnera, M.A., Li, R., Cai, L., Zhan, M., et al., 2011. Diagnosis of lung cancer in individuals with solitary pulmonary nodules by plasma microRNA biomarkers. BMC Cancer 11, 374.

Shersher, D.D., Vercillo, M.S., Fhied, C., Basu, S., Rouhi, O., Mahon, B., Coon, J.S., Warren, W.H., Faber, L.P., Hong, E., et al., 2011. Biomarkers of the insulin-like growth factor pathway predict progression and outcome in lung cancer. Ann. Thorac. Surg. 92, 1805–1811, discussion 1811.

Si, H., Sun, X., Chen, Y., Cao, Y., Chen, S., Wang, H., Hu, C., 2013. Circulating microRNA-92a and microRNA-21 as novel minimally invasive biomarkers for primary breast cancer. J. Cancer Res. Clin. Oncol. 139, 223–229.

Tan, A.R., Steinberg, S.M., Parr, A.L., Nguyen, D., Yang, S.X., 2008. Markers in the epidermal growth factor receptor pathway and skin toxicity during erlotinib treatment. Ann. Oncol. 19, 185–190.

Teng, H.-W., Wang, H.-W., Chen, W.-M., Chao, T.-C., Hsieh, Y.-Y., Hsih, C.-H., Tzeng, C.-H., Chen, P.C.-H., Yen, C.-C., 2011. Prevalence and prognostic influence of genomic changes of EGFR pathway markers in synovial sarcoma. J. Surg. Oncol. 103, 773–781.

Thompson, R.C., Herscovitch, M., Zhao, I., Ford, T.J., Gilmore, T.D., 2011. NF-kappaB down-regulates expression of the B-lymphoma marker CD10 through a miR-155/PU.1 pathway. J. Biol. Chem. 286, 1675–1682.

Thulin, P., Nordahl, G., Gry, M., Yimer, G., Aklillu, E., Makonnen, E., Aderaye, G., Lindquist, L., Mattsson, C.M., Ekblom, B., et al., 2014. Keratin-18 and microRNA-122 complement alanine aminotransferase as novel safety biomarkers for drug-induced liver injury in two human cohorts. Liver Int. 34, 367–378.

Ting, W.-C., Chen, L.-M., Pao, J.-B., Yang, Y.-P., You, B.-J., Chang, T.-Y., Lan, Y.-H., Lee, H.-Z., Bao, B.-Y., 2013. Common genetic variants in Wnt signaling pathway genes as potential prognostic biomarkers for colorectal cancer. PLoS One 8, e56196.

Tu, S., Jiang, H.W., Liu, C.X., Zhou, S.M., Tao, S.C., 2014. Protein microarrays for studies of drug mechanisms and biomarker discovery in the era of systems biology. Curr. Pharm. Des. 20, 49–55.

Wang, J., Yu, G., 2013. A systems biology approach to characterize biomarkers for blood stasis syndrome of unstable angina patients by integrating microRNA and messenger RNA expression profiling. Evid. Based Complement. Altern. Med. 2013, 510208.

Wang, J.-L., Hu, Y., Kong, X., Wang, Z.-H., Chen, H.-Y., Xu, J., Fang, J.-Y., 2013. Candidate microRNA biomarkers in human gastric cancer: a systematic review and validation study. PLoS One 8, e73683.

Wipfler, P., Heikkinen, A., Harrer, A., Pilz, G., Kunz, A., Golaszewski, S.M., Reuss, R., Oschmann, P., Kraus, J., 2013. Circadian rhythmicity of inflammatory serum parameters: a neglected issue in the search of biomarkers in multiple sclerosis. J. Neurol. 260, 221–227.

Yi, C., Mu, L., de la Longrais, I.A.R., Sochirca, O., Arisio, R., Yu, H., Hoffman, A.E., Zhu, Y., Katsaro, D., 2010. The circadian gene NPAS2 is a novel prognostic biomarker for breast cancer. Breast Cancer Res. Treat. 120, 663–669.

Yoon, S.O., Chun, S.-M., Han, E.H., Choi, J., Jang, S.J., Koh, S.A., Hwang, S., Yu, E., 2011. Deregulated expression of microRNA-221 with the potential for prognostic biomarkers in surgically resected hepatocellular carcinoma. Hum. Pathol. 42, 1391–1400.

Zhang, F., Chen, J.Y., 2010. Discovery of pathway biomarkers from coupled proteomics and systems biology methods. BMC Genom. (Suppl. 2), S12.

Zhu, X., Morales, F.C., Agarwal, N.K., Dogruluk, T., Gagea, M., Georgescu, M.-M., 2013. Moesin is a glioma progression marker that induces proliferation and Wnt/β-catenin pathway activation via interaction with CD44. Cancer Res. 73, 1142–1155.

Zhu, Y., Brown, H.N., Zhang, Y., Stevens, R.G., Zheng, T., 2005. Period3 structural variation: a circadian biomarker associated with breast cancer in young women. Cancer Epidemiol. Biomark. Prev. 14, 268–270.

CHAPTER SEVEN

Understanding Dynamical Diseases: Translational Bioinformatics Approaches

7.1 SPATIAL COMPLEXITY IN SYSTEMS BIOLOGY

As discussed in Chapters 1 and 2, the translation of pharmacogenomics and systems biology into personalized and systems medicine relies on the understanding of the complexity and dynamics in living systems. As illustrated in Fig. 7.1, such complexity can be dissected into the spatial and temporal dimensions to facilitate the translation. It is critical to examine and model the spatial dynamics at multiple length and time scales for understanding the complex biological systems and disease processes.

Considering the spatial dimension, the profiling for systems medicine should incorporate the information from different spatial levels including molecular, cellular, organ, psychosocial, and environmental interactions (Klann and Koeppl, 2012). Data and information for such systemic profiles may come from dynamical and integrative explorations based on the detections in various levels including nanotechnology and high-throughput (HTP) studies, biochemistry and biophysics analyses, as well as imaging and physiological assessments.

To elucidate the complexity at the molecular level, emphasis can be put on not only the genetic variations and polymorphisms but also the protein–protein interactions, as well as the real-time expression profiles (Huang and Wikswo, 2006). For the insights of the complexity at the cellular level, the measurements of cellular compartmentalization, dynamical protein complexes, and cellular network communications would be helpful. As an example, abnormal redox organization is one of the essential factors of diseases. The understanding of the redox compartmentalization in the process of cellular stress can be critical for revealing dynamical structure–function interactions to promote more effective preventive and therapeutic strategies (Jones and Go, 2010).

Translational Bioinformatics and Systems Biology Methods for Personalized Medicine
ISBN 978-0-12-804328-8
http://dx.doi.org/10.1016/B978-0-12-804328-8.00007-3

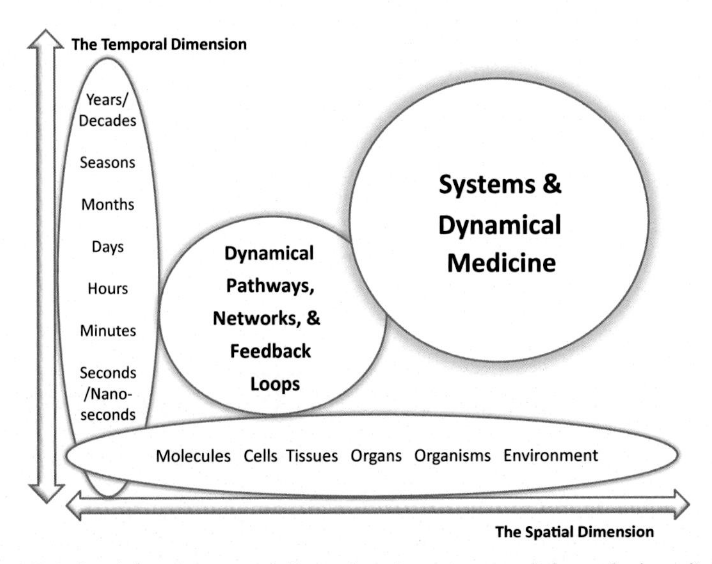

Figure 7.1 Spatial and temporal dimensions in systems and dynamical medicine.

Considering the spatial sizes, genes and proteins may be analyzed at the nanometer scale, and cellular organelles such as mitochondria may be examined at the micrometer scale (Huang and Wikswo, 2006). Animal or human tissues can be assessed at the millimeter scale, whereas the structures and functions of organs can be measured at the centimeter scale. In addition, the pathophysiology of the whole organisms may be detected at the meter level. Such systemic examinations may incorporate different physical parameters such as temperature, pressure, currents, and motions.

Another pivotal point for the systems-based analyses is to overcome the gaps and obstacles between various levels and scales. Multidimensional linkages including the genotype–phenotype associations would help to explain the collective "emergent" features of the hierarchical complexity to provide a more systemic view for medical practice (see Chapter 2). The connections of the dynamics and networks between different levels or scales are the key for the translation of pathophysiological mechanisms into better diagnostic biomarkers and therapeutic targets.

The heart is used as an example, which has dynamical activities at multiple scales during cardiac excitation. At the sub-millisecond or millisecond scales, the open and close of single ion channels can be detected (Qu et al., 2011). At the level of the whole cell, the collective activities of thousands of ion channels lead to an action potential. At the level of the whole heart and at the time scale of seconds, the electric impulses are shown as the synchronous contractions of the cardiac ventricles (Qu et al., 2011). Such collective activities can be recorded by the electrocardiogram.

Furthermore, at the centimeter level, the stability of the cardiac rhythms may be related to the propagation of wave fronts (Wilders and Jongsma, 1993). At the nanometer level, such processes may be associated with the activities of the ion channels and gap junctions. To have a more complete view of the cardiac activities, the comprehensive level and scale transcending perceptions are necessary to elucidate the spatiotemporal complexity in health and diseases.

7.2 TEMPORAL COMPLEXITY IN SYSTEMS BIOLOGY

As mentioned in Chapter 2, pharmacogenomics emphasizes the variances in human genomics, pathophysiology, and therapeutic reactions. Many factors may be involved in such individual variances, including both spatial and temporal differences. In conventional studies many efforts have been made to detect the spatial changes. However, the temporal changes have often been ignored in basic research or clinical observations.

For instance, until now epidemiologic investigations of illnesses have been mostly about the "average" spatial models such as the factors of locations and racial groups. However, in recent years the spatiotemporal risk factors have received more and more attentions (Zhang et al., 2011).

Such improvements have profound translational meanings because various systems such as the central nervous system are deeply affected by temporal factors such as the circadian rhythms (Iris, 2008; Kopec and Carew, 2013; Gulsuner et al., 2013). In many cases, the biophysical and biochemical activities may be decided by the temporal or chronological elements. The structure–function correlations and activities at certain time points may be distinctive from those at other time points.

For example, the activities in the aging process are shown as evolutionary and progressive changes in which the temporal factors are essential (see Chapter 12). These changes can be observed at various spatial levels including different physiological activities, cell cycles, and motility, as well as gene expression patterns during different ages (Manor and Lipsitz, 2013; Jonker et al., 2013; Zykovich et al., 2013).

As shown in Fig. 7.1, nonlinear examinations based on time series may lead to more accurate modeling of the complexity and dynamics across different spatial levels. For the temporal investigations, the time scales range from nanoseconds to minutes, from days to months, and from seasons to decades.

The activities of the ion channels can be used as an example. Their gating events may be detected at the temporal scale of microseconds (Huang and Wikswo, 2006). Moving up through the spatial level, the depolarization of the heart may be assessed at the temporal scale of milliseconds. Furthermore, the stability of the cardiac cycles may be evaluated at the temporal scale of seconds. The longevity and the aging processes of the whole organism may be observed at the temporal scale of gigaseconds.

As shown in these examples, the detailed physiological and pathological activities need to be observed within different rhythms and cycles to understand their systemic complexity and dynamics. Considering the cycles, those having 24-h patterns are called circadian rhythms. The cycles longer than 1 day or 24 h are in the range of infradian rhythms (Halberg et al., 2009; Lopes et al., 2013). The menstrual cycle is an example. Those with cycling patterns shorter than a day or 24 h are in the range of ultradian rhythms. Many physiological activities are within this range, including the neuron firing rates and heartbeat rhythms.

Among these cycles, the circadian rhythm has been relatively thoroughly covered in recent biomedical research (Lopes et al., 2013). The studies in circadian systems biology showed that disturbances in the circadian timing system may lead to molecular dysfunctions and serious pathologies such as cancer (Fuhr et al., 2015; also see Chapter 6). The interactive networks among the clock genes are essential in numerous molecular and cellular processes.

Furthermore, other types of cycles of patterns also deserve more attention. Besides these natural cycles or frequencies including biological and environmental patterns, the social cycles and events may have significant impacts on psychological or physiological health, including the cycles of weekly workdays, holiday stress, and the cycles of school years.

To connect the spatial and temporal factors, the temporal cycles may be studied in a wide range of spatial levels. As an example, the cyclic dynamic patterns have been detected in gene expressions in cell cycles and altered cellular redox conditions (Klevecz et al., 2008). The cell cycles have been suggested as an evolving process as proven by the genome-wide fluctuations observed at the transcript and protein levels.

In addition, the spatiotemporal dynamics and oscillations are the prominent features in mitochondrial physiology. These properties have been detected at various levels including transmembrane potentials, heart excitation waves, neural dynamics, cognition, working memory, and activities of bacteria (Qu et al., 2011; Kurz et al., 2010; Schultze-Kraft et al., 2011; Stephane et al., 2012; Lenz and Søgaard-Andersen, 2011).

The illustration of the temporal patterns and cyclic activities may enrich the understanding of the spatial properties. The description of the communal or "emergent" features across various spatiotemporal levels and scales would enable a more systemic and dynamic view (see Fig. 7.1; also see Chapter 2).

For instance, the spatiotemporal profiles can be established for various growth factor (GF) signaling in the process of memory formation (Kopec and Carew, 2013). Such comprehensive profiles of the interactive GF signaling networks may facilitate the modeling of the behavioral and structural plasticity to support more precise diagnosis and treatments for relevant diseases.

In summary, translational bioinformatics and systems biology strategies may integrate the HTP techniques, experimental explorations, and computational models to predict the outcomes of the disturbed networks and the pathologies at various system levels. The integrative and interdisciplinary methods would allow for chronotherapies and the restoration of the powerful timing system in many complex diseases.

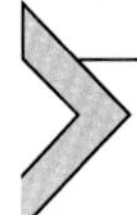

7.3 PROFILING OF DYNAMICAL DISEASES FOR SYSTEMS AND DYNAMICAL MEDICINE

As discussed earlier, the profiling of spatiotemporal patterns across various levels and scales from cells to organs and from seconds to days would allow for the translation of systems biology into systems and dynamical medicine. More integrative and personalized care would be possible by focusing on the systemic and dynamical changes with the elucidation of the nonlinearity and interrelationships of the complexity in health and diseases.

The emphasis on the term of "systems" in medicine would help to avoid the side effects caused by the drugs with single targets. The practice of "dynamical medicine" would embrace both of the spatial and temporal factors to understand the evolvement of health and diseases to improve the precision, prediction, and prevention in personalized medicine (see Chapter 2).

Mitochondria can be used as an example at the cellular level. The nonlinear dynamical activities of mitochondria have been associated with the energy homeostasis in liver cells (Ramanujan and Herman, 2007). The dysfunctions in these processes may be associated with aging and various illnesses including cardiovascular and metabolic diseases.

The heart can be used as an example at the organ level. The profiling of the nonlinear dynamics of heart rates may help address the complex elements for the diagnosis and therapy of relevant diseases. These elements include circadian patterns, the age factor, and the interactions with the autonomic nervous system (Vandeput et al., 2012).

The human nervous system is another example. It is composed of "a hierarchy of oscillatory processes" (Milton and Black, 1995). These dynamical activities interact extensively with various organs and systems, and have profound and bidirectional effects. The immune system is a good example (see Chapter 9).

The term "dynamical diseases" refers to the features of abnormal dynamical complexity and patterns in various disorders. The investigations of the features of dynamical diseases would elucidate the alterations and dysfunctions in the dynamics of the human body. Translational research of nonlinear dynamics may bring novel insights into the systems medicine of health and diseases.

The definition and focus on dynamical diseases may help in tracking the processes and patterns in which the symptoms appear and disappear over time with chaotic features (Bond and Guastello, 2013). The description of the dynamical and temporal patterns would enable the insight into the triggers for symptoms as well as the evolvement and progression of the disease processes as a whole (see Chapters 1 and 2).

Such studies may embrace the concepts in nonlinear dynamics including the stability and bifurcations of attractors (Glass, 2015). These features have been assessed in many diseases including depression, epilepsy, schizophrenia, substance abuse, Parkinson's disease, age-associated diseases, and hyperparathyroidism (Pezard et al., 1996; Schmid, 1991; an der Heiden, 2006; Lopes da Silva et al., 2003; Warren et al., 2003; Schiff, 2010; Edelstein-Keshet et al., 2001; Harms et al., 1992; also see Chapters 11 and 12).

For example, depression has been found to be featured with oscillating conditions in physiological and psychological aspects. Such patterns can be better described using the nonlinear parameters for the complex dynamics and networks (Tretter et al., 2011). In the case of obesity, the diurnal cortisol levels may be associated with the disorder in a nonlinear pattern (Kumari et al., 2010). In cancer, microRNA studies have identified the dynamical features and their close relationships with cell cycle regulations (Stahlhut Espinosa and Slack, 2006; also see Chapter 11). As a dynamical disease, chronic lymphocytic leukemia has been related to problems in B-cell cycles (Damle et al., 2010).

These facts demonstrate that "everything oscillates" (Klevecz et al., 2008) at various levels from genes to cells, from mitochondria to memory. The dynamical properties have been identified in all diseases, which should also be addressed in more accurate diagnosis and better targeted therapies. Systemic profiling and dynamical modeling need to emphasize the nonlinear time series discoveries, the rhythmic patterns, and the feedback networks to promote personalized medicine (Belair et al., 1995).

For instance, the complex temporal patterns related to biomarkers and symptoms have been studied in lung diseases including asthma (Frey et al., 2011). In psychiatric symptoms, the modeling of dynamical patterns may reflect the systematic variabilities (Odgers et al., 2009). In prostate cancer cells, the molecular elements that are involved in the treatment sensitivity and resistance have been featured with certain dynamical conditions and phases during the disease progression (Shaffer and Scher, 2003).

These evidences have indicated that the dynamical properties in the diseases need to be addressed with shifting targets at various levels including molecules and cellular networks during various stages for better therapies (see Chapter 8). In addition, the rhythmic and dynamical patterns of the diseases request efficient follow-ups to block the possible recurrence. For example, certain methods have been suggested for the prevention of recurrence among patients with depression (Pezard et al., 1996). These strategies would allow for the translation of dynamical systems biology into systems and dynamical medicine.

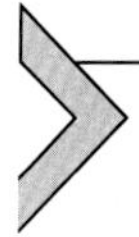

7.4 TRANSLATIONAL BIOINFORMATICS METHODS FOR STUDYING DYNAMICAL DISEASES

In addition to those discussed in Chapters 3 and 4, many databases and bioinformatics resources have been constructed to support translational studies in spatiotemporal dynamics. For example, the Systems Science of Biological Dynamics database (SSBD) (http://ssbd.qbic.riken.jp) is a database that can be applied for the analyses of quantitative data about spatiotemporal dynamics of biological phenomena (Tohsato et al., 2016).

SuperFly (http://superfly.crg.eu) provides an integrative platform for the examination of quantified spatiotemporal patterns in gene expressions in early dipteran embryos (Cicin-Sain et al., 2015). The Brain Transcriptome Database (BrainTx) (http://www.cdtdb.brain.riken.jp) may facilitate the profiling of spatiotemporal patterns in brain gene expressions (Sato et al.,

2008). It may support the visualization and assessment of transcriptome data associated with the development, function, and dysfunction phases and conditions of the brain.

Table 7.1 lists some examples of the translational bioinformatics and systems biology methods for the studies of dynamical diseases. The concept of "dynamical diseases" can be especially valuable for neuropsychiatry. The affective disorders show fluctuating state variables at both psychological and biological levels (Tretter et al., 2011). Data analyses at these systems levels have indicated the quasichaotic periodicity features of nonlinear dynamic systems. The systems-based approaches also address the inter- and intracellular networks and the dynamic cortisol regulation.

Such methodologies demonstrate that extremely complex diseases can be approached from analyzing simple nonlinear interactions among a few parameters. For instance, the theory of nonlinear dynamical systems was applied for the study of schizophrenia (an der Heiden, 2006). The study highlighted the feature of nonlinearity that the gradual changes of a single parameter (e.g., neurotransmitter dopamine, and serotonin) can lead to completely different types of behavior of the entire system (also see Chapter 2).

A mathematical model could be used to illustrate the activities of dopamine, including the excitatory–inhibitory circuits in the cortex, and the negative feedback loop among thalamus, prefrontal cortex, and striatum (an der Heiden, 2006). The models and concepts could represent the different patterns and the features of bifurcation, periodicities, and chaotic activities of dopamine. Here "bifurcation" refers to the phase transitions and the transitions between the different conditions.

In the case of depression, mathematical models with dynamical indicators can be applied to describe the brain activities with lower dynamical complexity in depressive patients (Pezard et al., 1996). Such analysis may differentiate patients with first episode from recurrent patients by highlighting their varied dynamical treatment responses. Such findings address the need for clinical follow-ups and certain interventions to prevent and treat the recurrence.

In obsessive–compulsive disorder (OCD), nonlinear regression parameters and dynamical studies may be helpful for the analysis of the intermittent outbursts of ritual behaviors (Bond and Guastello, 2013; also see Table 7.1). OCDs have shown a low-dimensional deterministic structure with significant rank order correlations. Such assessments may be used to understand the symptom severity and family reactions.

Table 7.1 Examples of Translational Bioinformatics and Systems Biology Methods for the Studies of Dynamical Diseases

Dynamical Diseases	Translational Bioinformatics Methods	References
Affective disorders, e.g., depression	• Fluctuating state variables on psychological and biological levels • Quasichaotic periodicity of non-linear dynamic systems • Inter- and intracellular networks	Tretter et al. (2011)
Colorectal cancer	• Immune cells interactions • Dynamical tumor microenvironment	Grizzi et al. (2013)
Depression	• Mathematical models with dynamical indicators • Dynamical complexity of brain activities • Dynamical treatment response differences between first episode and recurrent patients	Pezard et al. (1996)
Epilepsy	• Spatiotemporal assessments for the interactions in epileptic brain networks	Dickten et al. (2016)
Epilepsy	• The mathematics of nonlinear systems • Attractors about the trajectories with the initial status and outcomes • Computational models of neuronal networks for the simulations of neurophysiologic signals	Lopes da Silva et al. (2003)
Migraine	• Network studies • Mathematical modeling	Dahlem (2013)
Obsessive–compulsive disorder	• Nonlinear regression parameters • Rank order correlations	Bond and Guastello (2013)
Parkinson's disease	• Control theory • Computational neuroscience • Basal ganglia computational models	Schiff (2010)
Parkinson's disease	• Systemic variables • Models for negative feedbacks • Interactions between the central and peripheral loops, and other systemic signals	Beuter and Vasilakos (1995)

Continued

Table 7.1 Examples of Translational Bioinformatics and Systems Biology Methods for the Studies of Dynamical Diseases—cont'd

Dynamical Diseases	Translational Bioinformatics Methods	References
Schizophrenia	• Theory of nonlinear dynamical systems • A mathematical model • A negative feedback loop between thalamus, prefrontal cortex, and striatum • Analyses of patterns, bifurcation, periodicities, and chaotic activities	an der Heiden (2006)
Substance abuse	• Chaos theory and nonlinear dynamics • Nonlinearity in a time series of daily alcohol consumption	Warren et al. (2003)

The clinical observations of the time series of daily alcohol consumptions have pointed to the nonlinearity of substance abuse (Warren et al., 2003; also see Table 7.1). The chaos theory and nonlinear dynamics have been used to study this dynamical disease. The nonlinear model was found to be more accurate than a linear model. Such models would be more useful for the treatment of substance abuse and keeping sobriety.

Parkinson's disease is another example that is featured with short-term fluctuations in tremor. Other systemic variables include respiration and blood pressure (Beuter and Vasilakos, 1995). The model may address the negative feedback and the transient events, the interactions between the central and peripheral loops, and the interactions between the control loops and other systemic signals (see Table 7.1).

Such studies would emphasize the systemic variables and signals for representing the pathological features. In addition, the applications of control theory and computational neuroscience may help to build basal ganglia computational models (Schiff, 2010; also see Table 7.1). Such efforts may help in tracking observations.

Epilepsy is also a dynamical disease (Lopes da Silva et al., 2003). The concepts of the nonlinear systems such as the attractors may describe the trajectories with the initial status and outcomes. The computational models of neuronal networks have been found to be useful for the simulations of neurophysiologic signals (see Table 7.1). These analyses may identify the

EEG features before limbic seizures and the transition into paroxysmal epileptic activities.

In addition, the spatiotemporal assessments for the complex dynamics may help to describe the strength and direction of the interactions in the evolving large-scale epileptic brain networks (Dickten et al., 2016). The network theory may be useful for the illustration of both structure and function of epileptic networks. Such analyses can help to promote personalized diagnosis, treatment, and management.

Migraine is basically deemed as a dynamical disease where linear models have been found insufficient (Dahlem, 2013; also see Table 7.1). Minimally invasive and noninvasive neuromodulation techniques have been suggested for the potential solutions. The dynamical network studies may help to analyze the migraine generator region in the brainstem and facilitate mathematical modeling. The profiling of the migraine generator networks and spreading depression dynamics may be useful for identifying the neuromodulation targets in episodic migraine.

Colorectal cancer (CRC) is a multistep dynamical disease that may evolve over years through benign and malignant conditions, from single crypt lesions to malignant carcinoma with the possibilities for metastasis (Grizzi et al., 2013). Dynamical studies may help to understand the complex interactions among immune cells in the tumor microenvironment. The tumor stromal cells may also affect the growth and invasiveness of cancer cells in the dynamic tumor microenvironment. Such studies may help to understand the roles of both innate and adaptive immune cells in the local progression and metastasis. The dynamical features may contribute to the prognosis of CRC. More discussions in these aspects can be found in Chapter 11.

In summary, with the emphasis on the complex adaptive systems (CASs) and nonlinear dynamical features, systemic and dynamical profiling and modeling may help to solve the conceptual and technical difficulties. Such translational strategies may help to elucidate the disease complexity with the identification of various subtypes at various progressive phases (Hood and Flores, 2012). The identifications of alterations across various spatiotemporal levels and scales may improve the discovery of shifting treatment targets for individualized and timely interventions in personalized medicine (see Fig. 7.1). Dynamical and robust biomarkers may have critical roles in the stratification of disease subtypes and patient subgroups for optimized prevention and interventions.

REFERENCES

an der Heiden, U., 2006. Schizophrenia as a dynamical disease. Pharmacopsychiatry 39 (Suppl. 1), S36–S42.

Belair, J., Glass, L., An Der Heiden, U., Milton, J., 1995. Dynamical disease: identification, temporal aspects and treatment strategies of human illness. Chaos 5, 1–7.

Beuter, A., Vasilakos, K., 1995. Tremor: is Parkinson's disease a dynamical disease? Chaos 5, 35–42.

Bond, R.W., Guastello, S.J., 2013. Aperiodic deterministic structure of OCD and the familial effect on rituals. Nonlinear Dyn. Psychol. Life Sci. 17, 465–491.

Cicin-Sain, D., Pulido, A.H., Crombach, A., Wotton, K.R., Jiménez-Guri, E., Taly, J.-F., Roma, G., Jaeger, J., 2015. SuperFly: a comparative database for quantified spatio-temporal gene expression patterns in early dipteran embryos. Nucleic Acids Res. 43, D751–D755.

Dahlem, M.A., 2013. Migraine generator network and spreading depression dynamics as neuromodulation targets in episodic migraine. Chaos 23, 046101.

Damle, R.N., Calissano, C., Chiorazzi, N., 2010. Chronic lymphocytic leukaemia: a disease of activated monoclonal B cells. Best Pract. Res. Clin. Haematol. 23, 33–45.

Dickten, H., Porz, S., Elger, C.E., Lehnertz, K., 2016. Weighted and directed interactions in evolving large-scale epileptic brain networks. Sci. Rep. 6, 34824.

Edelstein-Keshet, L., Israel, A., Lansdorp, P., 2001. Modelling perspectives on aging: can mathematics help us stay young? J. Theor. Biol. 213, 509–525.

Frey, U., Maksym, G., Suki, B., 2011. Temporal complexity in clinical manifestations of lung disease. J. Appl. Physiol. 110, 1723–1731.

Fuhr, L., Abreu, M., Pett, P., Relógio, A., 2015. Circadian systems biology: when time matters. Comput. Struct. Biotechnol. J. 13, 417–426.

Glass, L., 2015. Dynamical disease: challenges for nonlinear dynamics and medicine. Chaos 25, 097603.

Grizzi, F., Bianchi, P., Malesci, A., Laghi, L., 2013. Prognostic value of innate and adaptive immunity in colorectal cancer. World J. Gastroenterol. 19, 174–184.

Gulsuner, S., Walsh, T., Watts, A.C., 2013. Spatial and temporal mapping of de novo mutations in schizophrenia to a fetal prefrontal cortical network. Cell 154, 518–529.

Halberg, F., Cornélissen, G., Wilson, D., et al., 2009. Chronobiology and chronomics: detecting and applying the cycles of nature. Biologist (London) 56, 209–214.

Harms, H.M., Prank, K., Brosa, U., et al., 1992. Classification of dynamical diseases by new mathematical tools: application of multi-dimensional phase space analyses to the pulsatile secretion of parathyroid hormone. Eur. J. Clin. Invest. 22, 371–377.

Hood, L., Flores, M., 2012. A personal view on systems medicine and the emergence of proactive P4 medicine: predictive, preventive, personalized and participatory. N. Biotechnol. 29, 613–624.

Huang, S., Wikswo, J., 2006. Dimensions of systems biology. Rev. Physiol. Biochem. Pharmacol. 157, 81–104.

Iris, F., 2008. Biological modeling in the discovery and validation of cognitive dysfunctions biomarkers. In: Turck, C.W. (Ed.), Biomarkers for Psychiatric Disorders. Springers Science + Business Media, New York.

Jones, D.P., Go, Y.-M., 2010. Redox compartmentalization and cellular stress. Diabetes Obes. Metab. 12 (Suppl. 2), 116–125.

Jonker, M.J., Melis, J.P.M., Kuiper, R.V., et al., 2013. Life spanning murine gene expression profiles in relation to chronological and pathological aging in multiple organs. Aging Cell 12, 901–909.

Klann, M., Koeppl, H., 2012. Spatial simulations in systems biology: from molecules to cells. Int. J. Mol. Sci. 13, 7798–7827.

Klevecz, R.R., Li, C.M., Marcus, I., et al., 2008. Collective behavior in gene regulation: the cell is an oscillator, the cell cycle a developmental process. FEBS J. 275, 2372–2384.
Kopec, A.M., Carew, T.J., 2013. Growth factor signaling and memory formation: temporal and spatial integration of a molecular network. Learn. Mem. 20, 531–539.
Kumari, M., Chandola, T., Brunner, E., et al., 2010. A nonlinear relationship of generalized and central obesity with diurnal cortisol secretion in the Whitehall II study. J. Clin. Endocrinol. Metab. 95, 4415–4423.
Kurz, F.T., Aon, M.A., O'Rourke, B., et al., 2010. Spatio-temporal oscillations of individual mitochondria in cardiac myocytes reveal modulation of synchronized mitochondrial clusters. Proc. Natl. Acad. Sci. U. S. A. 107, 14315–14320.
Lenz, P., Søgaard-Andersen, L., 2011. Temporal and spatial oscillations in bacteria. Nat. Rev. Microbiol. 9, 565–577.
Lopes Rda, S., Resende, N.M., Honorio-França, A.C., et al., 2013. Application of bioinformatics in chronobiology research. ScientificWorldJournal 2013, 153839.
Lopes da Silva, F., Blanes, W., Kalitzin, S.N., et al., 2003. Epilepsies as dynamical diseases of brain systems: basic models of the transition between normal and epileptic activity. Epilepsia 44 (Suppl. 12), 72–83.
Manor, B., Lipsitz, L.A., 2013. Physiologic complexity and aging: implications for physical function and rehabilitation. Prog. Neuropsychopharmacol. Biol. Psychiatry 45, 287–293.
Milton, J., Black, D., 1995. Dynamic diseases in neurology and psychiatry. Chaos 5, 8–13.
Odgers, C.L., Mulvey, E.P., Skeem, J.L., et al., 2009. Capturing the ebb and flow of psychiatric symptoms with dynamical systems models. Am. J. Psychiatry 166, 575–582.
Pezard, L., Nandrino, J.L., Renault, B., et al., 1996. Depression as a dynamical disease. Biol. Psychiatry 39, 991–999.
Qu, Z., Garfinkel, A., Weiss, J.N., Nivala, M., 2011. Multi-scale modeling in biology: how to bridge the gaps between scales? Prog. Biophys. Mol. Biol. 107, 21–31.
Ramanujan, V.K., Herman, B.A., 2007. Aging process modulates nonlinear dynamics in liver cell metabolism. J. Biol. Chem. 282, 19217–19226.
Sato, A., Sekine, Y., Saruta, C., Nishibe, H., Morita, N., Sato, Y., Sadakata, T., Shinoda, Y., Kojima, T., Furuichi, T., 2008. Cerebellar development transcriptome database (CDT-DB): profiling of spatio-temporal gene expression during the postnatal development of mouse cerebellum. Neural Netw. 21, 1056–1069.
Schiff, S.J., 2010. Towards model-based control of Parkinson's disease. Philos. Trans. A Math. Phys. Eng. Sci. 368, 2269–2308.
Schmid, G.B., 1991. Chaos theory and schizophrenia: elementary aspects. Psychopathology 24, 185–198.
Schultze-Kraft, M., Becker, R., Breakspear, M., et al., 2011. Exploiting the potential of three dimensional spatial wavelet analysis to explore nesting of temporal oscillations and spatial variance in simultaneous EEG-fMRI data. Prog. Biophys. Mol. Biol. 105, 67–79.
Shaffer, D.R., Scher, H.I., 2003. Prostate cancer: a dynamic illness with shifting targets. Lancet Oncol. 4, 407–414.
Stahlhut Espinosa, C.E., Slack, F.J., 2006. The role of microRNAs in cancer. Yale J. Biol. Med. 79, 131–140.
Stephane, M., Leuthold, A., Kuskowski, M., et al., 2012. The temporal, spatial, and frequency dimensions of neural oscillations associated with verbal working memory. Clin. EEG Neurosci. 43, 145–153.
Tohsato, Y., Ho, K.H.L., Kyoda, K., Onami, S., 2016. SSBD: a database of quantitative data of spatiotemporal dynamics of biological phenomena. Bioinformatics.
Tretter, F., Gebicke-Haerter, P.J., an der Heiden, et al., 2011. Affective disorders as complex dynamic diseases–a perspective from systems biology. Pharmacopsychiatry 44 (Suppl. 1), S2–S8.

Vandeput, S., Verheyden, B., Aubert, A.E., Van Huffel, S., 2012. Nonlinear heart rate dynamics: circadian profile and influence of age and gender. Med. Eng. Phys. 34, 108–117.
Warren, K., Hawkins, R.C., Sprott, J.C., 2003. Substance abuse as a dynamical disease: evidence and clinical implications of nonlinearity in a time series of daily alcohol consumption. Addict. Behav. 28, 369–374.
Wilders, R., Jongsma, H.J., 1993. Beating irregularity of single pacemaker cells isolated from the rabbit sinoatrial node. Biophys. J. 65, 2601–2613.
Zhang, Z., Chen, D., Liu, W., et al., 2011. Nonparametric evaluation of dynamic disease risk: a spatio-temporal kernel approach. PLoS One 6, e17381.
Zykovich, A., Hubbard, A., Flynn, J.M., et al., 2013. Genome-wide DNA methylation changes with age in disease free human skeletal muscle. Aging Cell.

PART THREE

Applications in Clinical and Translational Sciences

CHAPTER EIGHT

Translational Bioinformatics Methods for Drug Discovery and Development

8.1 CHALLENGES IN DRUG DISCOVERY AND POTENTIAL SOLUTIONS FROM PROFILING INTERACTOMES

An important application of systems biology and translational bioinformatics approaches is the discovery and development of novel and more effective drugs for complex diseases. The strategies of conventional drug discovery focus on the phenotypic and drug target screens and methods based on ligands and chemical structures (Prathipati and Mizuguchi, 2016). However, the reductionist views of "one-drug-fits-all" and "one gene–one disease–one drug" have led to the high toxicity, high rates of adverse events, and low effectiveness of many existing drugs.

Systems biology methods have been suggested to be useful for the identification and prediction of adverse drug reactions (ADRs). Conventional approaches such as clinical trials postmarket surveillance have drawbacks and limitations such as small sample size, biased analyses, with insufficient clinical evidences (Boland et al., 2016). The recent development in functional genomics and high-throughput (HTP) analyses may help in meeting the challenges by bringing better methodologies into the drug design processes (Prathipati and Mizuguchi, 2016).

As shown in Fig. 8.1, the establishment of systems-based profiles may reveal the comprehensive interactive networks and feedback loops for a better understanding of the efficacy and toxicity of different drugs. Specifically, the integration of different frameworks, resources, and multidimensional examinations would enable the modeling of the interrelationships among cellular networks, pathophysiological factors, and drug components. The applications of both experimental and computational methods in systems biology would allow for retrospective and prospective assessments for finding more effective and personalized medications (Prathipati and Mizuguchi, 2016).

Translational Bioinformatics and Systems Biology Methods for Personalized Medicine
ISBN 978-0-12-804328-8
http://dx.doi.org/10.1016/B978-0-12-804328-8.00008-5

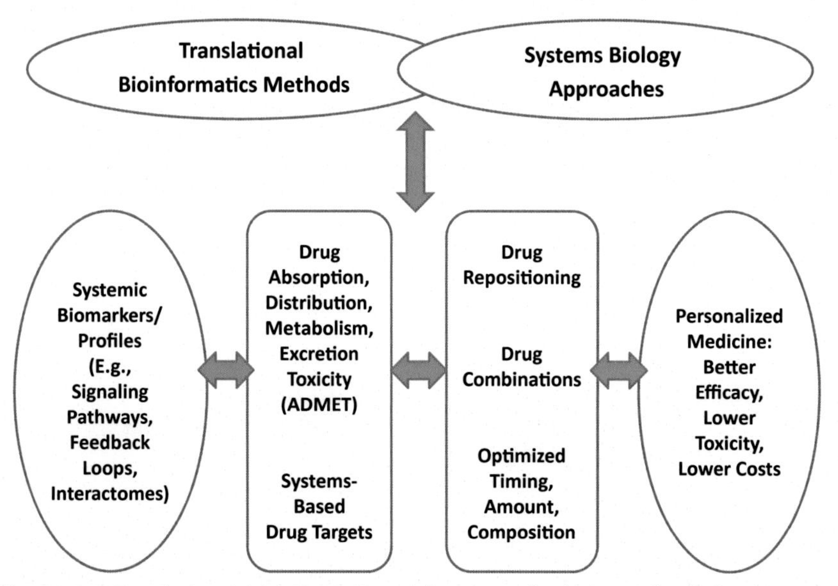

Figure 8.1 Translational bioinformatics and systems biology methods for systems-based drug discovery and personalized medicine.

Genes and proteins are performing in interactive networks and communicate multidirectionally with various cellular components. Such complex protein–protein interaction networks are defined as "interactomes" (Chautard et al., 2009). Systems biology strategies including HTP technologies and translational bioinformatics analyses of the multiparametric data sets can help to elucidate these interactomes to support more efficient drug targeting and development (see Fig. 8.1). Such strategies have been found to be useful for revealing interactive patterns in complex diseases including neurodegenerative diseases (NDs) (Vlasblom et al., 2014).

At the cellular level, the systems biology approaches based on high content screens would enable the examinations and detailed descriptions of the cellular alterations (Dunn et al., 2010). These descriptions can be integrated in transcriptomic and proteomic profiles, as well as the associations with phenotypic illnesses and drug reactions (see Chapter 3). Different levels of interactions and networks need to be integrated in the "interactomes" and systems-based profiles, from molecular genome to proteasome, from cellular mitochondrion to the whole organisms (Chautard et al., 2009).

For instance, the analyses of the dynamical activities and functions of the pivotal cellular organelle mitochondria may contribute to the better understanding of pathophysiological processes. The abnormal activities of mitochondria have been associated with many illnesses such as cancer, cardiovascular diseases, metabolic disorders, and NDs (Vlasblom et al., 2014). These properties would make the mitochondrial interactomes valuable for modeling the dynamical cellular networks and interactions. Such integrative models based on mitochondrial systems biology would contribute to the discovery of more comprehensive drug targets for NDs and other mitochondria-associated disorders.

In summary, translational bioinformatics has a critical role in analyzing such multifactorial, multiscale, and multidimensional data for the understanding and modeling of the complex networks and interactions. Various approaches can be used including principal component analysis (PCA) to reveal the complex patterns of the interrelationships (Dunn et al., 2010). The functional illustration of the cellular networks and systemic interconnections would allow for more efficient drug designs that go beyond the conventional approaches based on single molecules/targets to avoid the side effects and ADRs.

8.2 THE "TRANSLATIONAL" SIDE AND THE "BIOINFORMATICS" SIDE

As discussed earlier, approaches based on systems biology and translational bioinformatics such as the HTP methods and multilevel analyses of large data sets may help to construct the profiles for "interactomes" for finding novel drug targets. Such comprehensive drug targets would contribute to the better efficacy and lower expenses in the current costly drug development process (Vandamme et al., 2014; also see Fig. 8.1).

From the translational bioinformatics point of view, certain difficulties have to be overcome to promote more efficient drug discovery. One of such obstacles is the management and integration of various data types from the systems biology studies of multiple levels/scales (see Chapters 4 and 7).

For example, different entities and attributes need to be covered including various drugs and drug types, diseases and disease subtypes, and patients and patient subgroups. Another key component that has often been ignored is the interrelationships among these complex entities and attributes (Prathipati and Mizuguchi, 2016). Furthermore, more comprehensive data models, more effective data mining, and more efficient decision support tools need to be addressed to improve the current translational bioinformatics efforts.

Considering the "translational" side, the systems biology approaches can be addressed in the drug discovery efforts by embracing the network-based models with focuses on the molecular and cellular interactions that can be applied as the possible drug targets. This systems-based strategy may help with faster target validation and decreased attrition rates (Vandamme et al., 2014).

Such systems-based models and frameworks may be especially helpful for the discovery of multidrug treatments and drug combinations (see Fig. 8.1). By repositioning currently available drugs with the identification of patient subgroups, new combinations of existing drugs may promote the treatment efficacy and reduce the skyrocketing healthcare expenses.

Drug repositioning is the application of available drugs for treating conditions different from the original treatment purposes (Setoain et al., 2015). It is a very useful drug discovery tool that enables a quicker and inexpensive development process. Translational bioinformatics approaches can provide a dashboard to support repositioning hypotheses, to analyze transcriptomic data, and to identify biological associations between drugs and diseases. This is also the key step toward personalized medicine (see Fig. 8.1).

Complex disorders including cancer and cardiovascular diseases may benefit significantly from such systems-based models emphasizing the network interactions in the processes of cellular migration, proliferation, and drug resistance. By addressing mutations and escape pathways from multiple angles, drug combinations may be quite effective for solving the issues of drug resistance in these complex diseases (Ryall and Tan, 2015).

As an emerging interdisciplinary area, systems pharmacology may integrate chemical biology and heterogeneous data sources for the prediction of ADRs in individuals, groups, and global populations. The systems-based frameworks may embrace various data elements including diet and comorbidities for more accurate predictions (Boland et al., 2016).

Considering the "bioinformatics" side, a major task is to manage the huge amounts of the possible drug combinations and interactions (see Fig. 8.1). The benefits of bioinformatics strategies include the reduction of the searching space and time to promote the effectiveness of the time-consuming tests (Ryall and Tan, 2015). The combinations of systems biology modeling, HTP technologies, and translational bioinformatics strategies would facilitate better screening and decision-making for drug combinations.

In summary, systems biology and bioinformatics methodologies would embrace the multilevel modeling of functional genomics and signaling pathways (Ryall and Tan, 2015). Approaches including statistical association-based models would be useful for finding network-based biomarkers and "interactome" signatures for potential targets. These strategies may also enable computer-aided drug design and polypharmacology (Wathieu et al., 2016). The integration of the "translational" and "bioinformatics" aspects would enable the HTP screening processes for more efficient recognition of the optimal drug repositioning and combinations for personalized medicine (see Fig. 8.1).

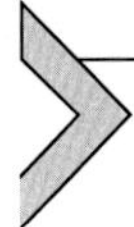

8.3 TRANSLATIONAL BIOINFORMATICS RESOURCES FOR DRUG DISCOVERY AND DEVELOPMENT

In addition to the resources discussed in Chapter 3, Table 8.1 lists some databases and bioinformatics resources that can be useful for translational studies in drug discovery and development. For instance, integrated database for ADMET and adverse effect predictive modeling (IDAAPM) is a comprehensive database about the absorption, distribution, metabolism, excretion, toxicity (ADMET), and adverse effects of drugs. It provides predictive modeling with the analysis of FDA drug data (Legehar et al., 2016).

The drug-minded protein interaction database (DrumPID) can be used to support target analysis and drug discovery (Kunz et al., 2016). CREDO collects data about structural interactomics to support drug development (Schreyer and Blundell, 2013). Protein–Drug Interaction Database (PDID) is a database about protein–drug interactions in the human proteome (Wang et al., 2016). ONRLDB is a manually curated database about ligands for orphan nuclear receptors that can be used for drug design (Nanduri et al., 2015).

The Mutations and Drugs Portal (MDP) is a platform connecting drug response data with genomic information (Taccioli et al., 2015). Virtually Aligned Matched Molecular Pairs Including Receptor Environment (VAMMPIRE) is a database about matched molecular pairs to support structure-based drug design and optimization (Weber et al., 2013; also see Table 8.1).

CancerDR is a database about cancer drug resistance (Kumar et al., 2013). ChEMBL is a large-scale bioactivity database that can be used to facilitate drug design (Gaulton et al., 2012). The Metabolism and Transport Drug Interaction Database is a platform to support the assessment of drug interaction evaluations (Hachad et al., 2010). PROMISCUOUS is a database that can be applied for network-based drug repositioning (von Eichborn et al., 2011; also see Table 8.1).

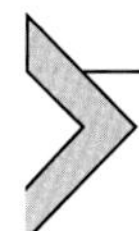

8.4 TRANSLATIONAL BIOINFORMATICS METHODS FOR DRUG DISCOVERY AND DEVELOPMENT

Most of the therapeutic strategies target the late stages in the pathological process with low predictive values and effective rates (Readhead and Dudley, 2013). However, the novel messages from studies on genomic

Table 8.1 Translational Bioinformatics Resources for Drug Discovery and Development

Databases/Tools	Web URL	Contents	References
CancerDR	http://crdd.osdd.net/raghava/cancerdr	Cancer drug resistance	Kumar et al. (2013)
ChEMBL	https://www.ebi.ac.uk/chembldb	A bioactivity database	Gaulton et al. (2012)
CREDO	http://marid.bioc.cam.ac.uk/credo	Interactomics for drug discovery	Schreyer and Blundell (2013)
DrumPID	http://drumpid.bioapps.biozentrum.uni-wuerzburg.de	Target analysis	Kunz et al. (2016)
IDAAPM	http://idaapm.helsinki.fi	ADMET and adverse effects with predictive modeling	Legehar et al. (2016)
Mutations and Drugs Portal (MDP)	http://mdp.unimore.it/	Linking drug response data to genomic information	Taccioli et al. (2015)
ONRLDB	http://www.onrldb.org	Validated ligands for orphan nuclear receptors	Nanduri et al. (2015)
Protein–Drug Interaction Database (PDID)	http://biomine.ece.ualberta.ca/PDID	Protein–drug interactions in the human proteome	Wang et al. (2016)
PROMISCUOUS	http://bioinformatics.charite.de/promiscuous	Network-based drug repositioning	von Eichborn et al. (2011)
The Metabolism and Transport Drug Interaction Database	http://www.druginteractioninfo.org	Drug interaction evaluation	Hachad et al. (2010)
VAMMPIRE	http://vammpire.pharmchem.uni-frankfurt.de	Structure-based drug design and optimization	Weber et al. (2013)

ADMET, absorption, distribution, metabolism, excretion, toxicity; *IDAAPM*, integrated database for ADMET and adverse effect predictive modeling.

and environmental interactions would enable more accurate and preventive strategies. Translational bioinformatics may have a pivotal role in this process by bridging the gap between the research data and clinical drug discovery toward the intent of personalized medicine.

Translational bioinformatics can provide powerful support for organizing and mining the data for network modeling, disease classification, biomarker discovery, and drug targeting and repositioning (see Fig. 8.1). Specifically, multiscale network disease models with high predictive values can be developed with the integration of information about gene expression, clinical traits, and other parameters. Causal network inference methods can be applied to identify the "key drivers" of pathology and the precise biomarker candidates of the etiology of diseases (Readhead and Dudley, 2013).

Note that in computational drug design and translational bioinformatics, data integration is especially important for the understanding of data from HTP technologies and studies on proteomics and transcriptomics (Seoane et al., 2013; also see Chapter 3). Data integration would allow for the access and queries to heterogeneous data sources (see Chapter 4). Approaches including federated databases, data warehouses, and semantic technology would facilitate information retrieval, clinical diagnosis, and drug discovery.

As translational medical strategies are gaining momentum in the biomedical society, translational bioinformatics may help to overcome the challenges in the pharmaceutical industry (Buchan et al., 2011). Table 8.2 shows some recent examples of translational bioinformatics methods for drug discovery and development.

For instance, integrative analyses were performed about the methylated genes associated with drug resistance in ovarian cancer (Yan et al., 2016; also see Table 8.2). Comprehensive bioinformatics examinations emphasized protein interactions, biological process enrichment, and annotations. The study showed a direct interaction between the phosphatase and tensin (PTEN) homolog gene and most of the other genes, pointing to the major regulatory roles of PTEN among these genes. The study highlighted the significance of the methylated genes in the regulation of resistant ovarian cancer. Such findings may be helpful for the prognosis of ovarian cancer.

Another example is about the off-label drug selections among triple negative breast cancer (TNBC) patients (Cheng et al., 2016; also see Table 8.2). A personalized medicine knowledge base was established with the integration of cancer drugs, drug target databases, and knowledge sources to support target selections. The analysis was performed

Table 8.2 Examples of Translational Bioinformatics Methods for Drug Discovery and Development

Associated Conditions	Translational Bioinformatics Methods	References
Anti-HIV/AIDS drugs and drug resistance	• Analysis of protein residues with digital signal processing • A drug resistance calculator	Nwankwo and Seker (2010)
Drug abuse and neuro-AIDS	• Databases to analyze molecular relationships • A public domain database	Shapshak et al. (2006)
Drug discovery and repositioning for IBD and autoimmune diseases	• HTP computations for large-scale data, genes and microRNAs • Clinically relevant gene-level profiling	Clark et al. (2012)
Drug repositioning	• Analysis of transcriptomic data for drug–disease relationships	Setoain et al. (2015)
Drug repositioning in organ transplantation	• Meta-analyses of genomic data and drugs • Finding redundant molecular pathways • Profiling of microarray data sets	Roedder et al. (2013)
Drug resistance in the treatment of ovarian cancer	• Analysis of the methylated genes associated with drug resistance • Analyses of protein interactions • Biological process enrichment	Yan et al. (2016)
Drug targets for osteoporosis	• Gene expression profiles from GEO • Functional pathway enrichment • GO and dysfunctional pathways • The connectivity map	Yu et al. (2013)
HCV drug discovery	• A knowledge discovery system for literature mining • Integration of the dictionary-based filtering and gene mention tagger	Seoud (2011)

Table 8.2 Examples of Translational Bioinformatics Methods for Drug Discovery and Development—cont'd

Associated Conditions	Translational Bioinformatics Methods	References
Glycomics and drug targets	• Data integration • Public databases for glycome informatics, e.g., KEGG • Tree-based models and algorithms for glycan structure data	Aoki-Kinoshita and Kanehisa (2006)
Off-label drug selection for TNBC	• The integration of databases of cancer drugs and drug targets • Analysis of the TNBC patient data • Analysis of personal molecular profiles	Cheng et al. (2016)

GEO, Gene Expression Omnibus; *GO*, gene ontology; *HCV*, Hepatitis C Virus; *HTP*, high-throughput; *IBD*, inflammatory bowel disease; *KEGG*, Kyoto Encyclopedia of Genes and Genomes; *TNBC*, triple negative breast cancer.

about TNBC patient data in the Cancer Genome Altar. A bioinformatics method was developed for cancer drugs selection by incorporating personal molecular profile data such as copy number variations, mutations, and gene expressions. The study identified some additional targets that had not been fully examined in the TNBC, such as Gamma-Glutamyl Hydrolase and Protein Tyrosine Kinase 6. Such translational bioinformatics approaches and knowledge bases may support the development of cancer precision medicine, including the off-label cancer drug applications in clinics.

In a study about drug targets for osteoporosis, the gene expression profiles of osteoporosis were constructed from the data in Gene Expression Omnibus (Yu et al., 2013; also see Table 8.2). The differentially expressed genes (DEGs) were analyzed using classical *t*-test method. The functional pathway enrichment analyses were applied to find the dysregulated gene ontology categories and dysfunctional pathways. The connectivity map was established to find compounds that induced inverse gene alterations. The study found that DEGs were enriched in nine pathways including the mitogen activated protein kinase (MAPK) signaling pathway. In addition, sanguinarine was found as a potential therapeutic drug candidate.

Translational bioinformatics approaches have also been found to be useful for the reposition of FDA-approved drugs in organ transplantation (Roedder et al., 2013; also see Table 8.2). The meta-analyses of genomic data and drug databases and a bioinformatics approach were applied for finding redundant molecular pathways. For example, significant enrichment was revealed for the IL-17 pathway. The methods included the analyses and profiling of microarray data sets from human renal allograft biopsies. These methods allow for a drug repositioning approach by using available drugs, which would lower the costs (see Fig. 8.1).

For inflammatory bowel disease (IBD) and other autoimmune diseases, translational bioinformatics methods may also contribute to drug discovery and repositioning of existing drugs (Clark et al., 2012; also see Table 8.2). Using bioinformatics tools and HTP computations for large-scale data, gene and microRNA biomarkers could be identified. The clinically relevant gene-level profiling of IBD subtypes and their association with autoimmune diseases may help to discover drug candidates for repositioning. The highly expressed IBD genes may become drug targets for gastrointestinal cancers, viral infections, and autoimmunity diseases including rheumatoid arthritis and asthma.

In the study for Hepatitis C Virus (HCV) drug discovery, a bioinformatics knowledge discovery system called BioHCVKD was established for literature mining and annotation of relevant HCV information (Seoud, 2011; also see Table 8.2). It integrated the dictionary-based filtering and conditional random field–based gene mention tagger. It was supported by the Abstract Insertion module and the Protein Insertion module. It may help to identify proteins, ligands, and active residues to support drug discovery.

In the assessment of anti-HIV/AIDS drugs and drug resistance, a signal processing-based bioinformatics method was applied for the examination of protein residues using digital signal processing techniques including informational spectrum method (ISM) (Nwankwo and Seker, 2010; also see Table 8.2). The methods integrated ISM, protein sequence information, and other relevant information. The digital approach to assess drug resistance can be used in other drug resistance studies to establish a computer-aided drug resistance calculator.

To study drug abuse and Neuro-AIDS, databases can be established to analyze the molecular relationships (Shapshak et al., 2006; also see Table 8.2). The investigation of gene expression interactions may help to explain the significance and complexity of the problems. The robust database systems may contain large data sets and serve as a public domain database for a shared platform to query, deposit, and review information.

In the analyses of glycans and diseases, translational bioinformatics approaches are critical for understanding glycomics and drug discovery (Aoki-Kinoshita and Kanehisa, 2006; also see Table 8.2). Various public databases can be used for glycome informatics including Kyoto Encyclopedia of Genes and Genomes (KEGG) GLYCAN, glycoSCIENCES.de, and the Consortium for Functional Glycomics. The tree-based models and algorithms have been found to be helpful to analyze glycan structure data. The integration of these data sets and informatics techniques may facilitate information extraction and support the discovery of biomarkers and drug targets.

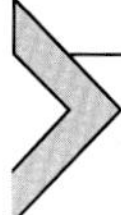

8.5 CONCLUSION: SYSTEMS-BASED MODELS AND DECISION SUPPORT FOR DRUG DISCOVERY

The applications of translational bioinformatics may serve as the pivotal bridge between the basic research in systems biology and clinical performance of systems and personalized medicine. The analyses of research data and patient records may support the patient-centered objectives and the decision-making in drug development pipelines in both research and clinical environments. With the support of dynamical, robust, and systems-based models for diseases and drug reactions, the drug discovery processes may be significantly improved with lower costs and higher effectiveness.

These integrative methods would contribute to the finding of more comprehensive drug targets and candidates. As mentioned earlier, conventional drug discovery approaches of structure-based targets for reducing unintentional binding may lead to adverse events caused by the unexpected multilevel interactions (Brown and Okuno, 2012). The systems-based profiling would help to identify multilevel variables as biomarkers for better drug targets, such as the interactions among proteins and metabolites in the pathophysiological networks (Galizzi et al., 2013).

These models may be especially helpful for the decision-making processes in drug discovery. They can be applied to examine the "what if?" scenarios in silico to predict the possible outcomes of the interventions to promote the efficiencies and reduce the expenses (Kell, 2013). These robust biomarkers embracing interactions and networks may represent the dynamical variances in personalized parameters.

The comprehensive models and decision-making tools would be especially helpful for embracing polypharmacology into the drug cocktails and combinations (Brown and Okuno, 2012; Kell, 2013). Such strategies would transform the conventional drug discovery routines from the single target into the "function-first" or phenotypic selecting methods emphasizing systemic networks (see Fig. 8.1). For instance, the comprehensive profiling of membrane transporters networks essential in drug absorption, distribution, metabolism, and excretion in different tissues may have broad implications in personalized medicine (Kell, 2013).

In summary, the integrative models developed from systems biology and translational bioinformatics strategies would contribute to the discovery of systems-based biomarkers and more effective treatments. Such approaches would enable the simulation, detection, and prediction of disease progression and drug responses for improving the safety, utilization, and effects among new and existing drugs.

REFERENCES

Aoki-Kinoshita, K.F., Kanehisa, M., 2006. Bioinformatics approaches in glycomics and drug discovery. Curr. Opin. Mol. Ther. 8, 514–520.

Boland, M.R., Jacunski, A., Lorberbaum, T., Romano, J.D., Moskovitch, R., Tatonetti, N.P., 2016. Systems biology approaches for identifying adverse drug reactions and elucidating their underlying biological mechanisms. Wiley Interdiscip. Rev. Syst. Biol. Med. 8, 104–122.

Brown, J.B., Okuno, Y., 2012. Systems biology and systems chemistry: new directions for drug discovery. Chem. Biol. 19, 23–28.

Buchan, N.S., Rajpal, D.K., Webster, Y., Alatorre, C., Gudivada, R.C., Zheng, C., Sanseau, P., Koehler, J., 2011. The role of translational bioinformatics in drug discovery. Drug Discov. Today 16, 426–434.

Chautard, E., Thierry-Mieg, N., Ricard-Blum, S., 2009. Interaction networks: from protein functions to drug discovery. A review. Pathol. Biol. 57, 324–333.

Cheng, L., Schneider, B.P., Li, L., 2016. A bioinformatics approach for precision medicine off-label drug drug selection among triple negative breast cancer patients. J. Am. Med. Inform. Assoc. 23, 741–749.

Clark, P.M., Dawany, N., Dampier, W., Byers, S.W., Pestell, R.G., Tozeren, A., 2012. Bioinformatics analysis reveals transcriptome and microRNA signatures and drug repositioning targets for IBD and other autoimmune diseases. Inflamm. Bowel Dis. 18, 2315–2333.

Dunn, D.A., Apanovitch, D., Follettie, M., He, T., Ryan, T., 2010. Taking a systems approach to the identification of novel therapeutic targets and biomarkers. Curr. Pharm. Biotechnol. 11, 721–734.

von Eichborn, J., Murgueitio, M.S., Dunkel, M., Koerner, S., Bourne, P.E., Preissner, R., 2011. PROMISCUOUS: a database for network-based drug-repositioning. Nucleic Acids Res. 39, D1060–D1066.

Galizzi, J.-P., Lockhart, B.P., Bril, A., 2013. Applying systems biology in drug discovery and development. Drug Metabol. Drug Interact. 28, 67–78.

Gaulton, A., Bellis, L.J., Bento, A.P., Chambers, J., Davies, M., Hersey, A., Light, Y., McGlinchey, S., Michalovich, D., Al-Lazikani, B., et al., 2012. ChEMBL: a large-scale bioactivity database for drug discovery. Nucleic Acids Res. 40, D1100–D1107.

Hachad, H., Ragueneau-Majlessi, I., Levy, R.H., 2010. A useful tool for drug interaction evaluation: the University of Washington Metabolism and Transport Drug Interaction Database. Hum. Genom. 5, 61–72.

Kell, D.B., 2013. Finding novel pharmaceuticals in the systems biology era using multiple effective drug targets, phenotypic screening and knowledge of transporters: where drug discovery went wrong and how to fix it. FEBS J. 280, 5957–5980.
Kumar, R., Chaudhary, K., Gupta, S., Singh, H., Kumar, S., Gautam, A., Kapoor, P., Raghava, G.P.S., 2013. CancerDR: cancer drug resistance database. Sci. Rep. 3, 1445.
Kunz, M., Liang, C., Nilla, S., Cecil, A., Dandekar, T., 2016. The drug-minded protein interaction database (DrumPID) for efficient target analysis and drug development. Database (Oxford) 2016.
Legehar, A., Xhaard, H., Ghemtio, L., 2016. IDAAPM: integrated database of ADMET and adverse effects of predictive modeling based on FDA approved drug data. J. Cheminform. 8, 33.
Nanduri, R., Bhutani, I., Somavarapu, A.K., Mahajan, S., Parkesh, R., Gupta, P., 2015. ONRLDB–manually curated database of experimentally validated ligands for orphan nuclear receptors: insights into new drug discovery. Database (Oxford) 2015.
Nwankwo, N., Seker, H., 2010. A signal processing-based bioinformatics approach to assessing drug resistance: human immunodeficiency virus as a case study. Conf. Proc. IEEE Eng. Med. Biol. Soc. 2010, 1836–1839.
Prathipati, P., Mizuguchi, K., 2016. Systems biology approaches to a rational drug discovery paradigm. Curr. Top. Med. Chem. 16, 1009–1025.
Readhead, B., Dudley, J., 2013. Translational bioinformatics approaches to drug development. Adv. Wound Care (New Rochelle) 2, 470–489.
Roedder, S., Kimura, N., Okamura, H., Hsieh, S.-C., Gong, Y., Sarwal, M.M., 2013. Significance and suppression of redundant IL17 responses in acute allograft rejection by bioinformatics based drug repositioning of fenofibrate. PLoS One 8, e56657.
Ryall, K.A., Tan, A.C., 2015. Systems biology approaches for advancing the discovery of effective drug combinations. J. Cheminform. 7, 7.
Schreyer, A.M., Blundell, T.L., 2013. CREDO: a structural interactomics database for drug discovery. Database (Oxford) 2013, bat049.
Seoane, J.A., Aguiar-Pulido, V., Munteanu, C.R., Rivero, D., Rabunal, J.R., Dorado, J., Pazos, A., 2013. Biomedical data integration in computational drug design and bioinformatics. Curr. Comput. Aided Drug Des. 9, 108–117.
Seoud, R.A., 2011. BioHCVKD: a bioinformatics knowledge discovery system for HCV drug discovery - identifying proteins, ligands and active residues, in biological literature. Int. J. Bioinform. Res. Appl. 7, 317–333.
Setoain, J., Franch, M., Martínez, M., Tabas-Madrid, D., Sorzano, C.O.S., Bakker, A., Gonzalez-Couto, E., Elvira, J., Pascual-Montano, A., 2015. NFFinder: an online bioinformatics tool for searching similar transcriptomics experiments in the context of drug repositioning. Nucleic Acids Res. 43, W193–W199.
Shapshak, P., Duncan, R., Turchan, J., Nath, A., Minagar, A., Kangueane, P., Davis, W., Chiappelli, F., Elkomy, F., Seth, R., et al., 2006. Bioinformatics models in drug abuse and neuro-AIDS: using and developing databases. Bioinformation 1, 86–88.
Taccioli, C., Sorrentino, G., Zannini, A., Caroli, J., Beneventano, D., Anderlucci, L., Lolli, M., Bicciato, S., Del Sal, G., 2015. MDP, a database linking drug response data to genomic information, identifies dasatinib and statins as a combinatorial strategy to inhibit YAP/TAZ in cancer cells. Oncotarget 6, 38854–38865.
Vandamme, D., Minke, B.A., Fitzmaurice, W., Kholodenko, B.N., Kolch, W., 2014. Systems biology-embedded target validation: improving efficacy in drug discovery. Wiley Interdiscip. Rev. Syst. Biol. Med. 6, 1–11.
Vlasblom, J., Jin, K., Kassir, S., Babu, M., 2014. Exploring mitochondrial system properties of neurodegenerative diseases through interactome mapping. J. Proteom. 100, 8–24.
Wang, C., Hu, G., Wang, K., Brylinski, M., Xie, L., Kurgan, L., 2016. PDID: database of molecular-level putative protein-drug interactions in the structural human proteome. Bioinformatics 32, 579–586.

Wathieu, H., Issa, N.T., Byers, S.W., Dakshanamurthy, S., 2016. Harnessing polypharmacology with computer-aided drug design and systems biology. Curr. Pharm. Des. 22, 3097–3108.

Weber, J., Achenbach, J., Moser, D., Proschak, E., 2013. VAMMPIRE: a matched molecular pairs database for structure-based drug design and optimization. J. Med. Chem. 56, 5203–5207.

Yan, B., Yin, F., Wang, Q.I., Zhang, W., Li, L.I., 2016. Integration and bioinformatics analysis of DNA-methylated genes associated with drug resistance in ovarian cancer. Oncol. Lett. 12, 157–166.

Yu, G., Wang, L., Li, Y., Ma, Z., Li, Y., 2013. Identification of drug candidate for osteoporosis by computational bioinformatics analysis of gene expression profile. Eur. J. Med. Res. 18, 5.

CHAPTER NINE

Translational Bioinformatics and Systems Biology for Understanding Inflammation

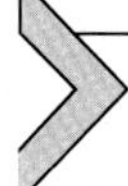

9.1 INTRODUCTION: SYSTEMS BIOLOGY, TRANSLATIONAL BIOINFORMATICS, AND INFLAMMATION

Systems biology studies have highlighted the roles of low-grade inflammation and metabolic factors in various complex conditions such as aging and age-associated diseases (Calçada et al., 2014). For instance, aging is an evolving process featured with systemic low-grade inflammation (or called "inflammaging") and progressive decline of metabolic functions (see Chapter 12).

Systems biology–based investigations may reveal the dynamical processes in the systemic inflammation and metabolic dysfunctions at multiple spatial levels (e.g., molecular, cellular, organ) and time scales (see Chapter 7). Computational modeling of the interactions among inflammatory and metabolic mediators may contribute to integrative strategies such as nutritional interventions (Calçada et al., 2014).

For example, systems biology investigations of adipose tissue metabolism showed its importance in energy homeostasis, inflammation, and complex interactions with other physiological systems (Manteiga et al., 2013). The metabolic pathways are critical for the uptake and hydrolysis of lipids via positive and negative feedback hubs containing protein kinases and nuclear receptors.

The cascade of metabolic and signaling pathways is essential for tissue remodeling and inflammatory responses through a self-reinforcing cycle (Manteiga et al., 2013). The complex adipose regulatory networks containing multiple cell types require systems strategies for comprehensive explorations toward a better understanding of complex diseases including diabetes, cancer, and cardiovascular diseases.

Translational systems biology strategies have been proposed as the rational and evidence-based method for understanding inflammation and

Translational Bioinformatics and Systems Biology Methods for Personalized Medicine
ISBN 978-0-12-804328-8
http://dx.doi.org/10.1016/B978-0-12-804328-8.00009-7

supporting clinical application of personalized and predictive medicine. Such methodologies may allow for the elucidation of the complex and nonlinear processes in inflammation that are the keys to many diseases including sepsis, traumatic brain injury, liver failure, and wound healing (Mi et al., 2010).

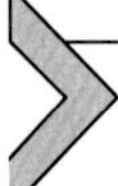

9.2 THE MICROBIOTA–GUT–BRAIN AXIS AND SYSTEMIC INFLAMMATION

In recent years increasing evidences have highlighted the significance of the microbiota–gut–brain (MGB) axis in health, inflammation, and diseases (Sherwin et al., 2016). The MGB axis is a bidirectional network between the brain and the gastrointestinal (GI) system with complex components involved such as the vagus nerve, immune factors, neuroendocrine pathways, and microbial metabolites (Sandhu et al., 2016). It can protect the host from detrimental pathogens and antigens. It can help with the digestion and metabolism of nutrients that are hard to access, including complex lipids and polysaccharides (Montiel-Castro et al., 2013).

This axis is essential for keeping homeostasis and normal behavior, especially cognitive processes including learning, memory, and decision-making mechanisms (Montiel-Castro et al., 2013). Dysfunctions in the axis have been associated with the alterations in neurotransmission and behavior, leading to various mental and metabolic disorders including anxiety, autism, depression, schizophrenia, obesity, cardiovascular diseases, and multiple sclerosis (MS) (Kennedy et al., 2016; Montiel-Castro et al., 2013). The MGB signaling may have profound implications about nonmotor symptoms in Parkinson's disease (Felice et al., 2016). The MGB axis is also critical for stress, visceral pain, and irritable bowel syndrome (Moloney et al., 2016).

Systems biology studies of the structural and functional dynamics in the MGB axis and relevant signaling pathways such as the kynurenine pathway may provide a pivotal model for the understanding of systemic inflammation and the translation from bench to bedside (Kennedy et al., 2016). Various spatiotemporal factors may be critical for the health of the MGB axis, such as genetics, mode of birth, nutritional status, and environment at different time points across the life span (Sandhu et al., 2016).

As an ecosystem, the human microbiota has many features of a complex adaptive system (CAS) including the diversity, dynamics, and adaptation (see Chapter 2). Inoculated before and after birth, the human microbiota is always in a coevolving process in the whole life span of the person to colonize, survive, and establish a mutualistic and symbiotic relationship with the host via finely tuned mechanisms (Mondot et al., 2013).

The human microbiota ecosystem regulates the immune function and influence the bowel motor mechanism (Kanauchi et al., 2013). In addition, it can neutralize carcinogens and drugs, and regulate intestinal motility. It enables visceral perception. In healthy conditions without perturbation, the human gut microbiota ecosystem keeps a globally stable state. This ecosystem is very resilient with adaptive features.

A neural network associated with the amygdala and the insular cortex may be involved in the integration of visceral inputs. The activation of the hypothalamic–pituitary–adrenal (HPA) axis may generate corticosteroids and regulate the composition of the gut microbiota. In addition, the neuronal efferent activation such as the "antiinflammatory cholinergic reflex" as well as sympathetic activation may lead to the release of neurotransmitters and influence the gut microbiota composition (Montiel-Castro et al., 2013).

Systems biology studies of the MGB axis may also help identify novel therapeutic targets for inflammation-associated diseases. Such examinations may reveal the bioactive microbial signals, the immune mediators, and relevant hormones associated with behavior and neurological functions (Elisei and de Castro, 2017). The adjustment of the gut microbiota back to its normal condition by using probiotics or prebiotics has been considered an important therapeutic choice (Kanauchi et al., 2013). Such intervention may be especially helpful for the treatment of the inflammation in the GI tract.

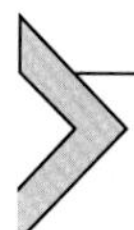

9.3 TRANSLATIONAL BIOINFORMATICS METHODS FOR THE STUDIES OF INFLAMMATION

In recent years more and more databases and bioinformatics resources have been developed for translational studies in inflammation and inflammation-related diseases. For example, GlycoGAIT (https://apps.connexios.com/glycogait) is a database about glycogenes and lectins associated with gastric inflammatory diseases (Oommen et al., 2016). IBDsite (http://www.itb.cnr.it/ibd/) provides a comprehensive platform to facilitate high-throughput data analysis about inflammatory bowel disease (IBD) (Merelli et al., 2012). Many of the resources discussed in Chapter 3 may also be applied.

As discussed previously, translational bioinformatics and systems biology would be the key for the integration, management, and mining of data to enable dynamical, explanatory, and predictive models (Voit, 2009). Such methods may provide a conceptual strategy for understanding the complex diseases in the context of inflammation and preconditioning. With the identification of high-dimensional biomarkers, the quantitative description of personalized health trajectories, and the construction of health risk profiles tracking the changes with age, better therapeutic options would be possible (see Chapters 2, 7, and 12).

Table 9.1 shows some examples of translational bioinformatics methods for studying inflammatory conditions. Data integration and data mining strategies are especially important for the translational efforts (see Chapter 4). Using a combination of proteomics, bioinformatics, and in silico interactomics, protein–protein interactions such as those between activated leukocytes and endothelial cells may become potential drug targets (Haqqani and Stanimirovic, 2013; also see Chapter 8). Such information can be applied for drug design such as neutralizing antibodies.

The bioinformatics workflow may also help identify cell–cell interactions to target certain inflammatory diseases and to improve currently available treatments. For example, literature mining and keyword searching were applied for finding genes and pathways implicated in radiation and immune/inflammatory responses in healthy and tumor tissues (Georgakilas et al., 2015). The shared genes in different phenomena may be involved in highly connected networks. These genes and pathways may be the potential biomarkers of responses to radiation with the underlying inflammatory mechanisms.

Chronic inflammatory processes are the features of cancers and liver cirrhosis. In the study of the evolution of inflammatory processes and liver cancer, bioinformatics and systems biology methods have been applied for the investigation of the complex cytokine networks or "cytokinome" (Costantini et al., 2014; Capone et al., 2014). The study of cytokinome may help elucidate the interactions among cytokines and other proteins in and around biological cells (see Table 9.1).

The construction of the cytokinome profiles may help reveal the complex interactions among cytokines, metabolic networks, natural antioxidants, inflammation, and liver cancer (Capone et al., 2014). Furthermore, such systems-based approaches may contribute to the investigations of adipokine interactome, obesity, type 2 diabetes, and chronic hepatitis C infection with the elucidation of the evolutionary processes in chronic inflammation (Costantini et al., 2014; Capone et al., 2014).

Table 9.1 Examples of Translational Bioinformatics and Systems Biology Methods for the Studies of Inflammation

Associated Conditions	Translational Bioinformatics Methods	References
Asthma	• Literature mining • Data meta-analysis • Combined gene-driven and pathway-driven strategies • Clinical and across-species data analysis	Riba et al. (2016)
Chronic inflammatory diseases, cancers, liver cirrhosis	• Cytokinome profiles • The complex interaction network of cytokines	Capone et al. (2014)
Inflammation	• Data integration and mining for dynamic, explanatory, and predictive models • A conceptual approach for formalizing information of inflammation	Voit (2009)
Inflammation, type 2 diabetes mellitus	• Multivariate statistical tools • A micronutrient phenotype database • Biological network models	van Ommen et al. (2008)
Inflammation and liver cancer	• Cytokinome profiles in the evolution of inflammatory processes • The complex interaction networks of cytokines • Adipokine interactome	Costantini et al. (2014)
Inflammatory disorders	• A combination of proteomics and in silico interactomics	Haqqani and Stanimirovic (2013)
Inflammatory responses and vaccination	• Systems vaccinology • Integrative analysis of innate and adaptive immunity in a quantitative framework • Blood transcriptomes	Zak and Aderem (2015)
Lung cancer	• Graph-based scoring function • Literature mining • Longitudinal proteomics analysis • Bioinformatics ranking algorithm	Oh et al. (2011)
Peripheral arterial disease	• Global protein–protein interaction networks of angiogenesis, immune responses, and arteriogenesis • Analyses of signaling pathways	Chu et al. (2015)
Tumors and inflammatory responses	• Literature mining • Keyword searching	Georgakilas et al. (2015)

In lung cancer, graph-based scoring mechanisms can be utilized to rank and identify robust biomarkers (Oh et al., 2011; also see Table 9.1). Literature mining and data analysis based on mass spectrometry may help assess the proximity between candidate proteins. Longitudinal proteomics and bioinformatics ranking algorithms may be used for discovering biomarkers from sample-limited clinical applications. The robust biomarkers can be applied for the detection of early signs, disease progression, and drug target selections.

For peripheral arterial disease (PAD), translational bioinformatics has been useful for the construction of global protein–protein interaction networks. These networks are critical in the systems biology studies of angiogenesis or "angiome" in "omics," immune responses or "immunome," and arteriogenesis or "arteriome" (Chu et al., 2015). With the mining of microarray gene expression data sets, possible drug targets and signaling pathways may be identified in the angiogenesis, immune, and arteriogenesis networks (see Table 9.1). For example, such assessments found the genes and pathways relevant to functional significance in PAD including TLR4, THBS1, and PRKAA2.

Because inflammation has a key role in type 2 diabetes mellitus, translational bioinformatics and systems biology strategies would facilitate multivariate analyses using nutrient-centered and physiology-centered parameters for the development of a micronutrient phenotype database (van Ommen et al., 2008). Such approaches would allow for the construction of network models embracing genes and protein–protein interactions for the identification of systems-based biomarkers representing target functions, biological responses, and metabolites (see Table 9.1). For instance, multiple micronutrients have been found critical in homeostasis and the prevention of chronic disorders, oxidation, and inflammation.

In the case of asthma, literature mining and data meta-analysis approaches were used to analyze microarray data sets (Riba et al., 2016). With the integration of gene-driven and pathway-driven strategies, translational bioinformatics may provide powerful tools for the analyses of clinical and across species data, as well as for the validation of conceptual and experimental models for finding new mechanisms (see Table 9.1). These approaches have led to the discovery of the significant roles of inflammation, circadian rhythms, peripheral genes, and superconnectors, for example, the pathways associated with IL-6, Stat1, Cadm1, and Erbb2.

Inflammatory responses are essential in vaccination. Systems vaccinology is the systems-based investigation of innate and adaptive immunity in a quantitative framework for the design and development of vaccines (Zak and Aderem, 2015). The combination of translational bioinformatics and systems biology strategies such as the analyses of blood transcriptomes and clinical data may help with the establishment of model systems to support the development of novel and more effective vaccines (see Table 9.1).

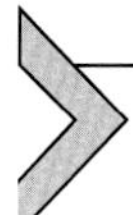

9.4 IDENTIFYING SYSTEMS-BASED BIOMARKERS FOR INFLAMMATION: EXAMPLES

9.4.1 Infectious Diseases

Similar to other illnesses, infectious diseases also have complex features with progressive stages and phases of development. For example, the disease caused by *Mycobacterium tuberculosis* may develop slowly because the relevant immune reactions and postexposure responses usually take several months to evolve.

In tuberculosis (TB), a broad spectrum of conditions and symptoms can be prompted by the infection, from asymptomatic infections at the beginning to severe tissue damages during later stages. The identification of the systems-based biomarkers needs to represent not only the evolutionary steps of the infections but also the host–immune responses and interactions to allow for the potential predictions and protections against the tissue damages and the overall illness.

Specifically, the clinical signs of *M. tuberculosis* infections may be closely associated with the evolution and progression of granulomatous lesions (Kunnath-Velayudhan and Gennaro, 2011). The granuloma structural alterations may be manifested in the peripheral circulation. Systems and dynamical examinations of biomarkers are needed to support accurate detection of the evolutionary progression of the disease. Such dynamical methodologies can help overcome the obstacles in detecting the progressive asymptomatic and symptomatic phases of infections to support more effective anti-TB preventions and interventions (Kunnath-Velayudhan and Gennaro, 2011).

For instance, an assessment of mycobacterium-induced cytokines including interferon gamma (IFN-γ), tumor necrosis factor alpha (TNF-α), interleukin 6 (IL-6), and IL-10 showed significant alterations of responses within different time frames (Talat et al., 2009). Such studies addressed the importance of analyzing multiple systems-based biomarkers in longitudinal studies to establish predictive biomarker profiles for more accurate diagnosis and prognosis of the disease.

On the basis of such analyses, infected individuals can be divided into asymptomatic and symptomatic groups, then subgroups at different stages of disease progression for better targeted and effective prevention and treatment. For example, the soluble members of the toll-like receptor 4 (TLR-4) pathway may be potential biomarkers to distinguish active TB from latent TB infection. The plasma levels of lipopolysaccharide and myeloid differentiation-2 have been suggested as the potential markers for the anti-TB therapeutic responses (Feruglio et al., 2013).

In patients with hepatitis C virus, components in the interferon signaling pathway such as IL28B and CXCL10 have been suggested as the prediction biomarkers for disease progression and the therapeutic responses to IFN-α2b/ribavirin (Helbig and Beard, 2012).

9.4.2 Inflammatory Bowel Disease/Crohn's Disease

Table 9.2 lists some examples of potential systems-based biomarkers in inflammation-associated diseases. A more complete and updated list can be found at the site of Biomarkers and Systems Medicine (BSM, 2016).

For example, the elements in the IL-23 signaling pathway especially the proteins downstream of IL-23 including regenerating protein 3β (REG), REG3γ, lipocalin 2 (LCN2), and macrophage migration inhibitory factor (MIF) have been found important in IBD/Crohn's disease (CD) (Cayatte et al., 2012; also see Table 9.2).

In addition, the components and interactions in the Wnt/β-catenin pathway such as β-catenin nuclear translocation, E-cadherin, and APC activities, as well as the expressions of c-Myc and Cyclin-D1 have also been suggested as the potential diagnostic and prognostic biomarkers for IBDs and sporadic colorectal cancer (Serafino et al., 2014).

9.4.3 Autoimmune Diseases

In MS, the members in the c-Jun N-terminal kinase (JNK)–dependent apoptosis pathway has been considered to have critical roles as the potential biomarkers for MS and relapsing-remitting MS (Ferrandi et al., 2011; also see Table 9.2). The relevant molecules include CD36, ITGAL, OLR1, PIAS-1, RTN4RL2, IL-23, and IFN-γ.

In rheumatoid arthritis (RA), the folate pathway including the levels of red blood cell methotrexate (MTX) and folate polyglutamate (PG) may have important roles (Dervieux et al., 2006). The genetic variants of the folate metabolic pathway enzymes including the gamma-glutamyl hydrolase gene (GGH) C-401T alleles have been suggested as the prediction biomarkers for therapeutic responses to the MTX therapy in RA (Hayashi et al., 2009).

Table 9.2 Examples of Potential Systems-Based Biomarkers in Inflammation-Associated Diseases

Associated Diseases	Potential Biomarkers	References
Active TB, LTBI, treatment responses	The toll-like receptor 4 (TLR-4) pathways	Feruglio et al. (2013)
Hepatitis C virus disease progression and treatment responses	The interferon signaling pathways (e.g., IL28B, CXCL10)	Helbig and Beard (2012)
IBD/CD	The IL-23 signaling pathway (e.g., proteins downstream of IL-23)	Cayatte et al. (2012)
IBD, SCC diagnostic and prognostic markers	The Wnt/β-catenin pathway components (e.g., E-cadherin, APC, c-Myc, Cyclin-D1)	Serafino et al. (2014)
IAR treatment	The glucocorticoid receptor pathways (e.g., CCL2, M-CSF, CXCL6, apoH)	Wang et al. (2011)
MS, RRMS active phases	The c-Jun N-terminal kinase (JNK)–dependent apoptosis pathways	Ferrandi et al. (2011)
RA (MTX effects and toxicity)	The folate pathway (e.g., MTX, folate PG)	Dervieux et al. (2006)
RA MTX treatment responses	The folate metabolic pathway enzyme polymorphisms (e.g., RFC G80A, GGH C-401T alleles)	Hayashi et al. (2009)

CD, Crohn's disease; *GGH*, gamma-glutamyl hydrolase gene; *IAR*, intermittent allergic rhinitis; *IBD*, inflammatory bowel disease; *LTBI*, latent TB infection; *MS*, multiple sclerosis; *MTX*, methotrexate; *PG*, polyglutamate; *RA*, rheumatoid arthritis; *RRMS*, relapsing-remitting MS; *SCC*, sporadic colorectal cancer; *TB*, Tuberculosis.

In allergic rhinitis, the components in the glucocorticoid (GC) receptor pathway and the acute phase response pathway including CCL2, M-CSF, and CXCL6 have been recommended as the potential therapeutic biomarkers among patients with intermittent allergic rhinitis under the GCs treatment (Wang et al., 2011; also see Table 9.2).

REFERENCES

BSM, 2016. Biomarkers and Systems Medicine. http://pharmtao.com/health/category/systems-medicine/biomarkers-systems-medicine.

Calçada, D., Vianello, D., Giampieri, E., Sala, C., Castellani, G., de Graaf, A., Kremer, B., van Ommen, B., Feskens, E., Santoro, A., et al., 2014. The role of low-grade inflammation and metabolic flexibility in aging and nutritional modulation thereof: a systems biology approach. Mech. Ageing Dev. 136–137, 138–147.

Capone, F., Guerriero, E., Colonna, G., Maio, P., Mangia, A., Castello, G., Costantini, S., 2014. Cytokinome profile evaluation in patients with hepatitis C virus infection. World J. Gastroenterol. 20, 9261–9269.

Cayatte, C., Joyce-Shaikh, B., Vega, F., Boniface, K., Grein, J., Murphy, E., Blumenschein, W.M., Chen, S., Malinao, M.-C., Basham, B., et al., 2012. Biomarkers of therapeutic response in the IL-23 pathway in inflammatory bowel disease. Clin. Transl. Gastroenterol. 3, e10.
Chu, L.-H., Vijay, C.G., Annex, B.H., Bader, J.S., Popel, A.S., 2015. PADPIN: protein-protein interaction networks of angiogenesis, arteriogenesis, and inflammation in peripheral arterial disease. Physiol. Genom. 47, 331–343.
Costantini, S., Colonna, G., Castello, G., 2014. A holistic approach to study the effects of natural antioxidants on inflammation and liver cancer. Cancer Treat. Res. 159, 311–323.
Dervieux, T., Greenstein, N., Kremer, J., 2006. Pharmacogenomic and metabolic biomarkers in the folate pathway and their association with methotrexate effects during dosage escalation in rheumatoid arthritis. Arthritis Rheum. 54, 3095–3103.
Elisei, C., de Castro, A.P., 2017. Insight into role of microbiota-gut-brain peptides as a target for biotechnology innovations. Front. Biosci. (Elite Ed.) 9, 76–88.
Felice, V.D., Quigley, E.M., Sullivan, A.M., O'Keeffe, G.W., O'Mahony, S.M., 2016. Microbiota-gut-brain signalling in Parkinson's disease: implications for non-motor symptoms. Parkinsonism Relat. Disord. 27, 1–8.
Ferrandi, C., Richard, F., Tavano, P., Hauben, E., Barbié, V., Gotteland, J.-P., Greco, B., Fortunato, M., Mariani, M.F., Furlan, R., et al., 2011. Characterization of immune cell subsets during the active phase of multiple sclerosis reveals disease and c-Jun N-terminal kinase pathway biomarkers. Mult. Scler. 17, 43–56.
Feruglio, S.L., Trøseid, M., Damås, J.K., Kvale, D., Dyrhol-Riise, A.M., 2013. Soluble markers of the Toll-like receptor 4 pathway differentiate between active and latent tuberculosis and are associated with treatment responses. PLoS One 8, e69896.
Georgakilas, A.G., Pavlopoulou, A., Louka, M., Nikitaki, Z., Vorgias, C.E., Bagos, P.G., Michalopoulos, I., 2015. Emerging molecular networks common in ionizing radiation, immune and inflammatory responses by employing bioinformatics approaches. Cancer Lett. 368, 164–172.
Haqqani, A.S., Stanimirovic, D.B., 2013. Prioritization of therapeutic targets of inflammation using proteomics, bioinformatics, and in silico cell-cell interactomics. Methods Mol. Biol. 1061, 345–360.
Hayashi, H., Fujimaki, C., Daimon, T., Tsuboi, S., Matsuyama, T., Itoh, K., 2009. Genetic polymorphisms in folate pathway enzymes as a possible marker for predicting the outcome of methotrexate therapy in Japanese patients with rheumatoid arthritis. J. Clin. Pharm. Ther. 34, 355–361.
Helbig, K.J., Beard, M.R., 2012. The interferon signaling pathway genes as biomarkers of hepatitis C virus disease progression and response to treatment. Biomark. Med. 6, 141–150.
Kanauchi, O., Andoh, A., Mitsuyama, K., 2013. Effects of the modulation of microbiota on the gastrointestinal immune system and bowel function. J. Agric. Food Chem. 61, 9977–9983.
Kennedy, P.J., Cryan, J.F., Dinan, T.G., Clarke, G., 2016. Kynurenine pathway metabolism and the microbiota-gut-brain axis. Neuropharmacology.
Kunnath-Velayudhan, S., Gennaro, M.L., 2011. Immunodiagnosis of tuberculosis: a dynamic view of biomarker discovery. Clin. Microbiol. Rev. 24, 792–805.
Manteiga, S., Choi, K., Jayaraman, A., Lee, K., 2013. Systems biology of adipose tissue metabolism: regulation of growth, signaling and inflammation. Wiley Interdiscip. Rev. Syst. Biol. Med. 5, 425–447.
Merelli, I., Viti, F., Milanesi, L., 2012. IBDsite: a galaxy-interacting, integrative database for supporting inflammatory bowel disease high throughput data analysis. BMC Bioinform. 13 (Suppl. 14), S5.
Mi, Q., Li, N.Y.-K., Ziraldo, C., Ghuma, A., Mikheev, M., Squires, R., Okonkwo, D.O., Verdolini-Abbott, K., Constantine, G., An, G., et al., 2010. Translational systems biology of inflammation: potential applications to personalized medicine. Per. Med. 7, 549–559.

Moloney, R.D., Johnson, A.C., O'Mahony, S.M., Dinan, T.G., Greenwood-Van Meerveld, B., Cryan, J.F., 2016. Stress and the microbiota-gut-brain axis in visceral pain: relevance to irritable bowel syndrome. CNS Neurosci. Ther. 22, 102–117.

Mondot, S., de Wouters, T., Doré, J., Lepage, P., 2013. The human gut microbiome and its dysfunctions. Dig. Dis. 31, 278–285.

Montiel-Castro, A.J., González-Cervantes, R.M., Bravo-Ruiseco, G., Pacheco-López, G., 2013. The microbiota-gut-brain axis: neurobehavioral correlates, health and sociality. Front. Integr. Neurosci. 7, 70.

Oh, J.H., Craft, J.M., Townsend, R., Deasy, J.O., Bradley, J.D., El Naqa, I., 2011. A bioinformatics approach for biomarker identification in radiation-induced lung inflammation from limited proteomics data. J. Proteome Res. 10, 1406–1415.

van Ommen, B., Fairweather-Tait, S., Freidig, A., Kardinaal, A., Scalbert, A., Wopereis, S., 2008. A network biology model of micronutrient related health. Br. J. Nutr. 99 (Suppl. 3), S72–S80.

Oommen, A.M., Somaiya, N., Vijayan, J., Kumar, S., Venkatachalam, S., Joshi, L., 2016. GlycoGAIT: a web database to browse glycogenes and lectins under gastric inflammatory diseases. J. Theor. Biol. 406, 93–98.

Riba, M., Garcia Manteiga, J.M., Bošnjak, B., Cittaro, D., Mikolka, P., Le, C., Epstein, M.M., Stupka, E., 2016. Revealing the acute asthma ignorome: characterization and validation of uninvestigated gene networks. Sci. Rep. 6, 24647.

Sandhu, K.V., Sherwin, E., Schellekens, H., Stanton, C., Dinan, T.G., Cryan, J.F., 2016. Feeding the microbiota-gut-brain axis: diet, microbiome, and neuropsychiatry. Transl. Res.

Serafino, A., Moroni, N., Zonfrillo, M., Andreola, F., Mercuri, L., Nicotera, G., Nunziata, J., Ricci, R., Antinori, A., Rasi, G., et al., 2014. WNT-pathway components as predictive markers useful for diagnosis, prevention and therapy in inflammatory bowel disease and sporadic colorectal cancer. Oncotarget 5, 978–992.

Sherwin, E., Sandhu, K.V., Dinan, T.G., Cryan, J.F., 2016. May the force be with you: the light and dark sides of the microbiota-gut-brain axis in neuropsychiatry. CNS Drugs 30, 1019–1041.

Talat, N., Shahid, F., Dawood, G., Hussain, R., 2009. Dynamic changes in biomarker profiles associated with clinical and subclinical tuberculosis in a high transmission setting: a four-year follow-up study. Scand. J. Immunol. 69, 537–546.

Voit, E.O., 2009. A systems-theoretical framework for health and disease: inflammation and preconditioning from an abstract modeling point of view. Math. Biosci. 217, 11–18.

Wang, H., Chavali, S., Mobini, R., Muraro, A., Barbon, F., Boldrin, D., Aberg, N., Benson, M., 2011. A pathway-based approach to find novel markers of local glucocorticoid treatment in intermittent allergic rhinitis. Allergy 66, 132–140.

Zak, D.E., Aderem, A., 2015. Systems integration of innate and adaptive immunity. Vaccine 33, 5241–5248.

CHAPTER TEN

Cardiovascular Diseases and Diabetes: Translational Bioinformatics and Systems Biology Methods

10.1 TRANSLATIONAL BIOINFORMATICS METHODS FOR STUDIES IN CARDIOVASCULAR DISEASES

In addition to those discussed in Chapters 3 and 4, many databases and bioinformatics resources can be useful for translational studies in cardiovascular diseases (CVDs). For example, CADgene (http://www.bioguo.org/CADgene/) is an integrative database about genes associated with coronary artery disease (CAD) (Liu et al., 2011). The Cardiac Atlas Project (http://www.cardiacatlas.org) is an imaging database about bioinformatics modeling and statistical atlases of the heart (Fonseca et al., 2011). LipidHome (http://www.ebi.ac.uk/metabolights/lipidhome/) is a platform that can be used to support the studies of lipids and lipidomics (Foster et al., 2013).

Table 10.1 shows some examples of translational bioinformatics methods for the studies of CVDs. For example, for CAD, genome-wide association studies (GWAS) were applied together with the assessment of population-specific linkage disequilibrium structures from 1000 Genomes Project (Bastami et al., 2016; also see Chapter 3). These approaches were applied to map disease correlations with microRNA (miRNA) targetome. The functional prediction strategies may reveal the effects of disease-related variants on miRNA targetome and their functional impacts. Such analyses may enable polymorphic miRNA targeting for more effective treatments.

In another study about CAD, gene expression profiles from the Gene Expression Omnibus (GEO) database were analyzed (Zhang et al., 2014b; also see Chapter 3). Statistical and bioinformatics approaches including database analyses were applied to examine the differentially expressed genes (DEGs) in CAD and the protein–protein interaction (PPI) networks for

Translational Bioinformatics and Systems Biology Methods for Personalized Medicine
ISBN 978-0-12-804328-8
http://dx.doi.org/10.1016/B978-0-12-804328-8.00010-3

Table 10.1 Examples of Translational Bioinformatics Methods for Cardiovascular Diseases

Associated Diseases	Translational Bioinformatics Methods	References
ACS	• DNA microarray data from ACS patients • GEO database analysis • DEGs by Affy packages of R • Interaction network analysis using STRING • DrugBank analysis for relevant small molecules	Zhang et al. (2014a)
CAD	• GWAS • Disease mapping to miRNA targetome • Functional prediction analyses to prioritize DAVs	Bastami et al. (2016)
CAD	• Gene expression profiles from the GEO database • Analysis of DEGs, PPI networks • Enriched biological processes among the DEGs using GO terms • Pathway analysis using the KEGG database	Zhang et al. (2014b)
CAD	• Integration of the genomics and proteomics data • Gene expression studies • Network modules	Vangala et al. (2013)
CVDs	• The informational spectrum method for structure/function analysis about lipoprotein lipase	Glisic et al. (2008)
CVDs	• Algorithms for the analysis of miRNAs • LocARNA and miRBase for structure analysis • Phylogenetic comparisons and RNA folding patterns • Functional target prediction • Analysis of signaling pathways	Kunz et al. (2015)
HIV-associated heart diseases	• Genome-wide proteomes at different stages of HIV replication and cell growth • Functional categorization and statistical analyses	Rasheed et al. (2015)

ACS, acute coronary syndrome; *CAD*, coronary artery disease; *CVDs*, cardiovascular diseases; *DAVs*, disease-associated variants; *DEGs*, differentially expressed genes; *GEO*, Gene Expression Omnibus; *GO*, Gene Ontology; *GWAS*, genome-wide association studies; *KEGG*, Kyoto Encyclopedia of Genes and Genomes; *miRNA*, microRNA; *PPI*, protein–protein interaction.

these DEGs. Integrative methods were used for the annotation and visualization to assess the enriched biological processes among the DEGs.

In addition, relevant pathways were collected from the information in Gene Ontology (GO) and the Kyoto Encyclopedia of Genes and Genomes (KEGG) pathway database (Zhang et al., 2014b; also see Chapter 3). Using the tools such as Cytoscape, the study examined the expression-activated subnetworks of the PPI networks and their topological features. The results of the study showed that the chemokine and focal adhesion signaling pathways may be important in the development of CAD.

In a study of acute coronary syndrome, DNA microarray data for thrombus-related leukocyte were examined for the associated genes (Zhang et al., 2014a). The study analyzed the data about associated small molecules from DrugBank. The study identified the significance of some upregulated genes from the chemokine family, including CCL2, CXCL3, and IL10, that were associated with the inflammatory activities. The identification of the inhibitors of CCL2 (L-Mimosine) from the DrugBank database may help support better drug targeting and more effective drug discovery.

As shown in these examples, translational bioinformatics and systems biology strategies have been found very useful for the integration of the genomics and proteomics data about CAD. For instance, the pathways associated with the core regulatory transcription factors, such as PPARG, EGR1, and ESRRA, may be potential biomarkers for the disease (Vangala et al., 2013). Such approaches for the development of multimarker modules and pathway-based biomarkers may be helpful for the predictions of disease risks.

Systems biology methods focusing on proteomics help revealed the cardiovascular extracellular matrix (ECM) as a hallmark for various CVDs (Barallobre-Barreiro et al., 2016). For instance, cardiac inflammation and ECM remodeling of myocardial infarction (MI) were identified as the significant components in the response of the left ventricle (Ghasemi et al., 2014). Data integration and mining of the data from high-throughput (HTP) genomic and proteomic studies extracted temporal and spatial information for the development of dynamical models to predict cardiac healing post-MI and to identify biomarkers (see Chapters 2 and 4).

The informational spectrum method (ISM) is a virtual spectroscopy approach in bioinformatics for the assessments of structures and functions of nucleotide and protein sequences. Using ISM in CVDs, the evolutionary information was examined for the structure of lipoprotein lipase (LPL) (Glisic et al., 2008). Mutations were found to change the LPL enzymatic activities. The bioinformatics approach for analyzing the pathogenic effect of LPL nonsynonymous single-nucleotide substitution may help recognize the risk factors and potential biomarkers for CVDs.

The combination of translational bioinformatics together with validation and screening experiments has been found necessary for the investigation of the complex interactions of miRNAs with the genome in CVDs (Kunz et al., 2015). The bioinformatics algorithms were applied for the studies of miRNAs and their regulatory elements in cardiac miRNA biology for the remodeling of their systemic effects. By using resources such as LocARNA and miRBase (see Chapter 3), the biogenesis of miRNAs and phylogenetic patterns may be revealed. Such strategies can also be used to explore the RNA folding patterns and signaling pathways for diagnostic biomarkers and therapeutic target predictions.

For the study of HIV-associated heart diseases, bioinformatics and statistical methods were used to investigate the genome-wide proteomes of a $CD4^{+}$ T-cell line during various stages of HIV replication and cell growth (Rasheed et al., 2015). The functional categorization identified multiple pathways such as those associated with the myosin light-chain kinase. These proteins and pathways may contribute to cardiac stress, arrhythmia, as well as cardiomyopathy and heart failure. Such translational efforts have been suggested useful for finding novel biomarkers in addition to the conventional markers to support the early diagnosis and more specific treatments.

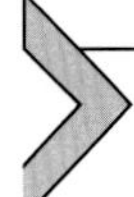

10.2 LIPIDOMICS, COMPUTATIONAL SYSTEMS BIOLOGY, AND DRUG REPOSITIONING

Atherosclerosis is a complex disease with evolving stages of inflammation and the hardening of the arterial wall. Various reasons may lead to the inflammatory status such as increased cellularity and cellular debris, lipid buildup, and the accumulation of extracellular materials (De Leon et al., 2015). In these cases, lipids have the pivotal roles because they are the essential elements of the vascular plaques (Ekroos et al., 2010). Studies using mass spectrometry examined and quantified hundreds of different molecular lipid species to identify the structure–function correlations (De Leon et al., 2015; Ekroos et al., 2010).

The term "lipidomics" refers to the "omics" and systems biology research about lipids, which may be critical in these efforts. The identifications of lipid- and lipidomics-based biomarkers may contribute to more accurate CVD diagnosis with better targeted treatments (Ekroos et al., 2010). Such approaches would be meaningful for the establishment of translational models to promote patient stratification and personalized therapeutic efficacy and safety.

Computational systems biology strategies have been proven useful for supporting drug repositioning in the treatment of CVDs. For example, bioinformatics methods were applied to screen for pathways in a cellular model with the integration of drug–transcriptome–response data sets and disease-associated genes for the mappings between drugs target pathways (Yu and Ramsey, 2016). The gene set enrichment analysis (GSEA) tests about drug targets and transcriptome profiles of atherosclerosis identified potential CVD targets such as those relevant to PPARγ and δ-opioid receptor. The data integration and mining approaches were found effective for the translational screening (see Chapter 4).

In summary, systems biology methods can be powerful for the translation of laboratory-based discoveries into safe and effective clinical outcomes. The integration of the information about the chemical structure and "omics" has been found useful for the examinations of drug effects for inflammation in atherosclerosis (Kleemann et al., 2011). Using approaches such as comparative genome-wide pathway mapping, predictive models can be constructed for identifying the potential drug targets, drug-responsive cellular pathways, and combination therapies. For example, the inflammatory signaling cascades relevant to cytokines, such as IFNγ and IL1β, and transcriptional regulators, such as NFκB and STAT3, may be especially important (Kleemann et al., 2011).

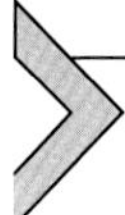

10.3 NUTRITIONAL SYSTEMS BIOLOGY, BIOMARKERS, AND TYPE 2 DIABETES

As a critical risk factor for cardiovascular morbidity and mortality, the networks in the macrovascular complications of diabetes may be essential in the disease onset and progression. The integrative proteomic and bioinformatic assessments of data from aortic vessels in diabetic models indicated the alterations of molecules and pathways associated with vascularization, hypertrophy, and amino acid breakdown in the development of atherosclerosis (Husi et al., 2014).

The systems biology based investigations in obesity, diabetes, and CVDs have also revealed the etiological activities in multiple disease-associated cells, tissues, and organs (Meng et al., 2013). Approaches using functional genomics, causality inference, and network development may contribute to the finding of biomarkers and possible drug targets from the integration of large-scale data sets.

Various "omics" branches may contribute to such purposes, including genomics, proteomics, transcriptomics, epigenomics, metabolomics, and microbiomics (see Chapter 3). The integration of computational, experimental, and clinical studies in translational bioinformatics and systems biology may facilitate the description of the dynamical activities and systemic understanding of the pathophysiological processes for personalized interventions.

For example, systems biology studies about metabolomics of obesity and type 2 diabetes (T2D) may be especially helpful for the detections of disease progression and potential biomarkers such as fatty acids and bile acids (Abu Bakar et al., 2015). The proteome-based systems biology examinations of the diabetic mouse aorta revealed the alterations in fatty acid biosynthesis as the potential hallmark for diabetes-related vascular disease (Husi et al., 2014).

Data mining approaches may be used to examine the multifaceted data sets from proteomic and HTP transcriptomic microarrays for analyzing the complex networks such as nutritional interactions (Moore and Weeks, 2011). Such research on nutritional systems biology may help elucidate the complex interrelationships among dietary nutrients, molecular and cellular elements, as well as phenotypic tissues, organs, systems, and diseases (Zhao et al., 2015). The findings about the disease-disturbed nutritional networks would contribute to the recognition of the nutritional biomarkers and systemic targets for better preventive and treatment methods (Moore and Weeks, 2011).

For example, genetic research has emphasized the critical roles of nutritional and dietary imbalances in the T2D pathogenesis and risks. High-fat diet (HFD) can be harmful to the normal functions at many levels from metabolites to microbiota, from genetic pathways to the NAD+/NADH ratio (Zhao et al., 2015). The dysfunctions of metabolites may result in altered DNA methylation and abnormal gene expressions. Such changes in epigenomics can cause transcriptional aberrant and the disturbance of circadian rhythms with the abnormal functions of the CLOCK/BMAL1 complex and PPARγ (Zhao et al., 2015).

Moreover, HFD can also lead to the decreased levels of butyrate-generating bacteria in the gut microbiota (see Chapter 9). HFD can lower the levels of short-chain fatty acids, such as butyrate, and cause the dysfunctions of histone and chromatin. The accumulation of these alterations including dysfunctions in the AMPK- and SIRT1-related signaling pathways may ultimately result in the malfunctions of mitochondria (Zhao et al., 2015). These factors highlight the importance of the systems-based profiling for T2D and CADs.

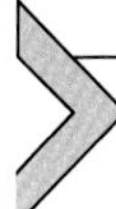

10.4 FINDING SYSTEMS-BASED BIOMARKERS FOR CARDIOVASCULAR DISEASES: EXAMPLES

Table 10.2 shows some examples of the potential systems-based biomarkers for CVDs. A more complete and updated list can be found at the site of Biomarkers and Systems Medicine (BSM, 2016).

For example, multiple pathways, such as the osteoprotegerin (OPG), receptor activator of nuclear factor-κB (RANK), and RANK ligand (RANKL) pathways, have been closely related to CVDs. As shown in Table 10.2, these pathways have been suggested as the potential predictive biomarkers for the risk factors, especially the CVD burden and mortality (Lieb et al., 2010).

The inflammatory pathways are also important. For example, the IL-33/ST2 pathway has been related to the intramyocardial

Table 10.2 Examples of Potential Systems-Based Biomarkers for Cardiovascular Diseases and Diabetes

Associated Conditions	Potential Biomarkers	References
CVD prediction	The OPG/RANK/RANKL pathways	Lieb et al. (2010)
CVDs, heart failure therapeutic targets	The IL-33/ST2 pathways	Kakkar and Lee (2008)
Differentiating ischemic from hemorrhagic stroke	The S100B/RAGE pathways	Montaner et al. (2012)
Sepsis-induced nonovert DIC diagnosis	TFPI and P-selectin	Mosad et al. (2011)
Sporadic thoracic aortic aneurysm in women	The TGF-β pathway gene polymorphisms	Scola et al. (2014)
Type 2 diabetes and subclinical atherosclerosis	Lipoxygenase pathway gene variations (e.g., polymorphisms in ALOX12, ALOX5, and ALOX5AP)	Burdon et al. (2010)

CVDs, cardiovascular diseases; *DIC*, disseminated intravascular coagulation; *OPG*, osteoprotegerin; *RANK*, receptor activator of nuclear factor-κB; *RANKL*, RANK ligand; *TFPI*, tissue factor pathway inhibitor.

fibroblast–cardiomyocyte interactions. This pathway has been recommended as a potential biomarker and treatment target for CVDs, especially heart failure (Kakkar and Lee, 2008; also see Table 10.2).

In addition, the plasma levels of the members relevant to the pathways of tissue factor pathway inhibitor (TFPI) and P-selectin may be potential diagnostic biomarkers for sepsis-induced nonovert disseminated intravascular coagulation (DIC) (Mosad et al., 2011; also see Table 10.2). The S100B/RAGE pathway has been indicated as the potential plasma biomarkers for differentiating ischemic from hemorrhagic stroke (Montaner et al., 2012). Genetic variants such as the polymorphisms [rs900 transforming growth factor beta 2 (TGF-β2) single-nucleotide polymorphism (SNP)] in the TGF-β pathway have been proposed as the possible biomarkers of sporadic thoracic aortic aneurysm in women (Scola et al., 2014).

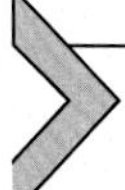

10.5 FINDING SYSTEMS-BASED DYNAMICAL BIOMARKERS FOR DIABETES: EXAMPLES

Studies using mice models have discovered two dynamical networks as the potential biomarkers to represent two important progression phases in type 1 diabetes (T1D) (Liu et al., 2013). The two phases were tightly related to periinsulitis and hyperglycemia. These dynamical network-based biomarkers may have important predictive values and serve as the early-warning sign of T1D for the prevention of disease progression.

For type 2 diabetes mellitus (T2DM), the approaches for finding dynamical network biomarkers (DNBs) may also be helpful. For instance, the tissue-specific DNBs and abnormal gene expressions have been discovered in the phases of T2DM transition and progression in different tissues, including the adipose, liver, and muscle (Li et al., 2014). These phases are crucial for insulin resistance and inflammation.

As another example, genetic variances of the members in the lipoxygenase pathway including the polymorphisms of ALOX12, ALOX5, and ALOX5AP have been closely associated with T2D and subclinical atherosclerosis. The biomarkers of inflammation (e.g., CRP, ICAM-1) and calcification (MGP) may also be important (Burdon et al., 2010; also see Table 10.2).

In summary, the major alterations in the relevant signaling pathways have been found especially important for the disease pathogenesis. The systems-based biomarkers such as the DNBs in the T2DM cases may be very helpful for the detection of the signs and emergence of the transition states for the early and immediate diagnosis and prognosis of the complex illnesses.

REFERENCES

Abu Bakar, M.H., Sarmidi, M.R., Cheng, K.-K., Ali Khan, A., Suan, C.L., Zaman Huri, H., Yaakob, H., 2015. Metabolomics – the complementary field in systems biology: a review on obesity and type 2 diabetes. Mol. Biosyst. 11, 1742–1774.

Barallobre-Barreiro, J., Lynch, M., Yin, X., Mayr, M., 2016. Systems biology – opportunities and challenges: the application of proteomics to study the cardiovascular extracellular matrix. Cardiovasc. Res. 112, 626–636.

Bastami, M., Nariman-Saleh-Fam, Z., Saadatian, Z., Nariman-Saleh-Fam, L., Omrani, M.D., Ghaderian, S.M.H., Masotti, A., 2016. The miRNA targetome of coronary artery disease is perturbed by functional polymorphisms identified and prioritized by in-depth bioinformatics analyses exploiting genome-wide association studies. Gene 594, 74–81.

BSM, K.P., Rudock, M.E., Lehtinen, A.B., Langefeld, C.D., Bowden, D.W., Register, T.C., Liu, Y., Freedman, B.I., Carr, J.J., Hedrick, C.C., 2016. Biomarkers and Systems Medicine. Mediators Inflamm. 2010, 170153. http://pharmtao.com/health/category/systems-medicine/biomarkers-systems-medicine.

Burdon, K.P., Rudock, M.E., Lehtinen, A.B., Langefeld, C.D., Bowden, D.W., Register, T.C., Liu, Y., Freedman, B.I., Carr, J.J., Hedrick, C.C., et al., 2010. Human lipoxygenase pathway gene variation and association with markers of subclinical atherosclerosis in the diabetes heart study. Mediators Inflamm. 2010, 170153.

De Leon, H., Boue, S., Szostak, J., Peitsch, M.C., Hoeng, J., 2015. Systems biology research into cardiovascular disease: contributions of lipidomics-based approaches to biomarker discovery. Curr. Drug Discov. Technol. 12, 129–154.

Ekroos, K., Jänis, M., Tarasov, K., Hurme, R., Laaksonen, R., Cowan, B.R., Dinov, I.D., Finn, J.P., Hunter, P.J., Kadish, A.H., 2010. Lipidomics: a tool for studies of atherosclerosis. Curr. Atheroscler. Rep. 12, 273–281.

Fonseca, C.G., Backhaus, M., Bluemke, D.A., Britten, R.D., Chung, J.D., Cowan, B.R., Dinov, I.D., Finn, J.P., Hunter, P.J., Kadish, A.H., et al., 2011. The cardiac atlas project–an imaging database for computational modeling and statistical atlases of the heart. Bioinformatics 27, 2288–2295.

Foster, J.M., Moreno, P., Fabregat, A., Jin, Y.-F., et al., 2013. LipidHome: a database of theoretical lipids optimized for high throughput mass spectrometry lipidomics. PLoS One 8, e61951–91.

Ghasemi, O., Ma, Y., Lindsey, M.L., Jin, Y.-F., Prljic, J., Veljkovic, N., 2014. Using systems biology approaches to understand cardiac inflammation and extracellular matrix remodeling in the setting of myocardial infarction. Wiley Interdiscip. Rev. Syst. Biol. Med. 6, 77–91.

Glisic, S., Arrigo, P., Alavantic, D., Perovic, V., Prljic, J., Veljkovic, N., Delles, C., Perco, P., Mischak, H., 2008. Lipoprotein lipase: a bioinformatics criterion for assessment of mutations as a risk factor for cardiovascular disease. Proteins 70, 855–862.

Husi, H., Van Agtmael, T., Mullen, W., Bahlmann, F.H., Schanstra, J.P., Vlahou, A., Delles, C., Perco, P., Mischak, H., 2014. Proteome-based systems biology analysis of the diabetic mouse aorta reveals major changes in fatty acid biosynthesis as potential hallmark in diabetes mellitus-associated vascular disease. Circ. Cardiovasc. Genet. 7, 161–170.

Kakkar, R., Lee, R.T., Perlina, A., Kaput, J., Verschuren, L., Wielinga, P.Y., Hurt-Camejo, E., Nikolsky, Y., van Ommen, B., Kooistra, T., 2008. The IL-33/ST2 pathway: therapeutic target and novel biomarker. Nat. Rev. Drug Discov. 7, 827–840.

Kleemann, R., Bureeva, S., Perlina, A., Kaput, J., Verschuren, L., Wielinga, P.Y., Hurt-Camejo, E., Nikolsky, Y., van Ommen, B., Kooistra, T., 2011. A systems biology strategy for predicting similarities and differences of drug effects: evidence for drug-specific modulation of inflammation in atherosclerosis. BMC Syst. Biol. 5, 125.

Kunz, M., Xiao, K., Liang, C., Viereck, J., Pachel, C., Frantz, S., Thum, T., Dandekar, T., Hoffmann, U., Fox, C.S., 2015. Bioinformatics of cardiovascular miRNA biology. J. Mol. Cell. Cardiol. 89, 3–10.

Li, M., Zeng, T., Liu, R., Chen, L., 2014. Detecting tissue-specific early warning signals for complex diseases based on dynamical network biomarkers: study of type 2 diabetes by cross-tissue analysis. Brief. Bioinform. 15, 229–243.

Lieb, W., Gona, P., Larson, M.G., Massaro, J.M., Lipinska, I., Keaney, J.F., Rong, J., Corey, D., Hoffmann, U., Fox, C.S., et al., 2010. Biomarkers of the osteoprotegerin pathway: clinical correlates, subclinical disease, incident cardiovascular disease, and mortality. Arterioscler. Thromb. Vasc. Biol. 30, 1849–1854.

Liu, H., Liu, W., Liao, Y., Cheng, L., Liu, Q., Ren, X., Shi, L., Tu, X., Wang, Q.K., Guo, A.-Y., 2011. CADgene: a comprehensive database for coronary artery disease genes. Nucleic Acids Res. 39, D991–D996.

Liu, X., Liu, R., Zhao, X.M., Chen, L., 2013. Detecting early-warning signals of type 1 diabetes and its leading biomolecular networks by dynamical network biomarkers. BMC Med. Genom. (Suppl. 2), S8.

Meng, Q., Mäkinen, V.-P., Luk, H., Yang, X., Giralt, D., Merino, C., Ribó, M., Rosell, A., Penalba, A., Fernández-Cadenas, I., 2013. Systems biology approaches and applications in obesity, diabetes, and cardiovascular diseases. Curr. Cardiovasc. Risk Rep. 7, 73–83.

Montaner, J., Mendioroz, M., Delgado, P., García-Berrocoso, T., Giralt, D., Merino, C., Ribó, M., Rosell, A., Penalba, A., Fernández-Cadenas, I., et al., 2012. Differentiating ischemic from hemorrhagic stroke using plasma biomarkers: the S100B/RAGE pathway. J. Proteom. 75, 4758–4765.

Moore, J.B., Weeks, M.E., Eltayeb, A.A., 2011. Proteomics and systems biology: current and future applications in the nutritional sciences. Adv. Nutr. 2, 355–364.

Mosad, E., Elsayh, K.I., Eltayeb, A.A., 2011. Tissue factor pathway inhibitor and P-selectin as markers of sepsis-induced non-overt disseminated intravascular coagulopathy. Clin. Appl. Thromb. Hemost. 17, 80–87.

Rasheed, S., Hashim, R., Yan, J.S., Bova, M., Forte, G.I., Pisano, C., Candore, G., Colonna-Romano, G., Lio, D., Ruvolo, G., 2015. Possible biomarkers for the early detection of HIV-associated heart diseases: a proteomics and bioinformatics predictionββ. Comput. Struct. Biotechnol. J. 13, 145–152.

Scola, L., Di Maggio, F.M., Vaccarino, L., Bova, M., Forte, G.I., Pisano, C., Candore, G., Colonna-Romano, G., Lio, D., Ruvolo, G., et al., 2014. Role of TGF-β pathway polymorphisms in sporadic thoracic aortic aneurysm: rs900 TGF-β2 is a marker of differential gender susceptibility. Mediat. Inflamm. 2014, 165758.

Vangala, R.K., Ravindran, V., Ghatge, M., Shanker, J., Arvind, P., Bindu, H., Shekar, M., Rao, V.S., 2013. Integrative bioinformatics analysis of genomic and proteomic approaches to understand the transcriptional regulatory program in coronary artery disease pathways. PLoS One 8, e57193Yu, A.Z., Ramsey, S.A., Liu, H., Zheng, C., Rao, K., Fang, Y., Zhou, H., Xiong, S., 2016. A computational systems biology approach for identifying candidate drugs for repositioning for cardiovascular disease. Interdiscip. Sci. 34, 863–869. http://dx.doi.org/10.1007/s12539-016-0194-3.

Zhang, L., Li, J., Liang, A., Liu, Y., Deng, B., Wang, H., 2014a. Immune-related chemotactic factors were found in acute coronary syndromes by bioinformatics. Mol. Biol. Rep. 41, 4389–4395.

Zhang, X., Cheng, X., Liu, H., Zheng, C., Rao, K., Fang, Y., Zhou, H., Xiong, S., 2014b. Identification of key genes and crucial modules associated with coronary artery disease by bioinformatics analysis. Int. J. Mol. Med. 34, 863–869.

Zhao, Y., Barrere-Cain, R.E., Yang, X., 2015. Nutritional systems biology of type 2 diabetes. Genes Nutr. 10, 481.

CHAPTER ELEVEN

Translational Bioinformatics and Systems Biology for Cancer Precision Medicine

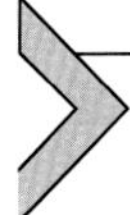

11.1 INTRODUCTION: SYSTEMS BIOLOGY, CANCER PRECISION MEDICINE, AND IMMUNOTHERAPY

Cancers are very complex illnesses caused by heterogeneous reasons at various system levels. For instance, breast cancer can be classified into different subtypes with different patterns of treatment responses and clinical outcomes (Yersal and Barutca, 2014). However, conventional grouping methods are not able to provide accurate and comprehensive classifications for diagnosis and prognosis. The objectives of personalized medicine for cancer therapies request novel strategies that are systems based and practical in the clinics.

The heterogeneous complexity in cancers refer to a broad spectrum of mechanisms, especially that different genotypes may be related to similar clinical phenotypes (Yersal and Barutca, 2014). Such mechanisms refer to the importance of the biomarkers based on genomics findings and interactive networks to enable better diagnosis and therapeutic target selections, as well as the prevention of disease development or recurrence.

For example, in the processes of breast tumor metastasis, the tumor proliferative capacity is a key element. Systems-based biomarkers such as the survival-associated subnetworks may be useful for representing the tumor proliferative potential (Song et al., 2015). These comprehensive networks may include profiles at various system levels, such as the gene expression patterns and protein–protein interaction networks. The systems-based profiling of cancer genes, cellular pathways, the dynamics of tumor metastases, and therapeutic outcomes may contribute to the discovery of more robust biomarkers with high predictive and prognostic values for the selection of treatment targets.

Translational Bioinformatics and Systems Biology Methods for Personalized Medicine
ISBN 978-0-12-804328-8
http://dx.doi.org/10.1016/B978-0-12-804328-8.00011-5

The personalized medicine models may replace the reductionist "one-size-fits-all" concept. The novel approaches for systems-based biomarker identification may also transform oncology into the biomarker-driven cancer precision medicine (Aftimos et al., 2014). Translational studies in Cancer Precision Medicine (CPM) should include topics such as cancer screening, the monitoring and prediction of relapse and recurrence, drug selection, drug response prediction, and personalized immunotherapy (Deng and Nakamura, 2016).

Various systems biology approaches may be applied for biomarker identifications. For instance, the immunohistochemistry (IHC) techniques have been useful for detecting protein expressions as the biomarkers for analyzing therapeutic responses in patients with solid tumors (Aftimos et al., 2014). The analyses in functional genomics can help identify different subtypes of genetic abnormalities, including the copy number alterations and sequence mutations. The comprehensive profiling containing information about the "gene expression signatures" and cellular networks may contribute to the discovery of novel anticancer targets for individualized adjuvant treatment in various tumor types.

Specifically, the immune functions are crucial in cancer pathophysiology and therapeutic responses. The immune system may interact with and regulate the growth and development of tumors at various molecular, cellular, tissue, and organ levels. On the basis of such understanding, a new strategy of immunotherapy emerged with the significant potentials as the effective treatment for cancers. In recent years more and more applications of immunotherapies have been approved by the US FDA (Guhathakurta et al., 2013). However, these novel methods need to be improved to increase the efficacy, reduce the expenses, and limit the potential adverse events.

To achieve these goals, systems biology and "omics"-based strategies such as the high-throughput (HTP) profiling are necessary to elucidate the complex mechanisms underlying tumor immunosurveillance and different immune phenotypes. Integrative translational studies of the immunological activities at different system levels can be applied such as microarrays, deep sequencing, and mass spectrometry (MS) (Guhathakurta et al., 2013).

Because multiple pathways are altered in tumors, especially those involved in redox and immune regulations, they should be considered as systems-based biomarkers that can be useful for the identification of extracellular and intracellular therapeutic targets. For example, thioredoxin 1 (Trx1) is an important redox regulator, while CD30 is a crucial cell membrane receptor involved in immune responses (Berghella et al., 2011). The CD30/Trx1 system plays significant roles in immune homeostasis and can be a potential target for cancer therapy toward the concurrent optimization of both the redox and the immune regulations.

Translational bioinformatics methods would be especially important in analyzing the tremendous amount of data from these comprehensive analyses. Methods such as data mining and decision support would contribute to the development of systems-based models for the discovery of diagnostic and prognostic biomarkers, target selections, and outcome analysis (see Chapter 4). Such translational efforts would be helpful for improving cancer immunotherapies.

For example, in a study for precision cancer prognosis, data mining strategies were developed for examining big genomics and clinical data (Ow and Kuznetsov, 2016). An algorithm for Prognostic Signature Vector Matching and multivariate prognostic models were developed based on methods including machine learning, K-nearest neighbor, random forest, neural networks, and logistic regression. The study recognized the potential prognostic significant mRNAs and the age factor to group ovarian cancer patients. The systems-based translational bioinformatics methods enabled the method for more precise and reproducible patient classification and disease predictions.

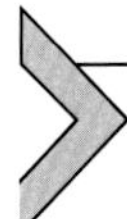

11.2 TRANSLATIONAL BIOINFORMATICS RESOURCES FOR CANCER STUDIES

Many resources are available for systems biology and translational studies in cancer. Some of these resources and methodologies have been discussed in Chapters 3–5. Table 11.1 lists some databases and bioinformatics resources that have been developed in recent years, which can be especially useful for translational studies in cancer.

For example, The Cancer Genome Atlas (TCGA) is an integrative platform supporting the studies in cancer genomics (Cancer Genome Atlas Research Network et al., 2013). The cBioPortal for Cancer Genomics is a portal for visualization, analysis, and download of cancer genomics data sets (Gao et al., 2013). SurvExpress is a database for biomarker validation based on the collection of cancer gene expression data (Aguirre-Gamboa et al., 2013).

The Precision Medicine Knowledgebase (PMKB) is an interactive platform to support the studies of structured clinical-grade cancer mutations and cancer precision medicine (Huang et al., 2016). The CancerResource

Table 11.1 Translational Bioinformatics Resources for Cancer Studies

Tools	Web URL	Contents	References
BcCluster	http://www.bccluster.org	Molecular signatures of bladder cancer	Bhat et al. (2016)
BCNTB	http://breastcancertissuebank.org/bioinformatics	Breast cancer tissue bank	Cutts et al. (2015)
Cancer Proteomics database	http://cancerproteomics.uio.no	Proteomic, prostate cancer and anticancer drugs	Arntzen et al. (2015)
CancerProView	http://cancerproview.dmb.med.keio.ac.jp/php/cpv.html	Graphical images of cancer-related genes and proteins	Mitsuyama and Shimizu (2012)
CancerResource	http://data-analysis.charite.de/care/	Drug-target relationships in cancer	Gohlke et al. (2016)
cBioPortal for Cancer Genomics	http://www.cbioportal.org/	Cancer genomics data sets	Gao et al. (2013)
miREC	http://www.mirecdb.org	miRNAs in endometrial cancer	Ulfenborg et al. (2015)
Mouse Tumor Biology (MTB)	http://tumor.informatics.jax.org/mtbwi/index.do	Mouse models	Bult et al. (2015)
Mutations and Drugs Portal (MDP)	http://mdp.unimore.it	Linking drug response data to genomic information	Taccioli et al. (2015)
Pancreatic Cancer Database	http://www.pancreaticcancerdatabase.org	Pancreatic cancer, changes at the mRNA, protein, and miRNA levels	Thomas et al. (2014)
Precision Medicine Knowledgebase (PMKB)	https://pmkb.weill.cornell.edu	Clinical-grade cancer mutations	Huang et al. (2016)
Prospective Lynch Syndrome Database	http://lscarisk.org	Risks for first cancer, subsequent cancer with Lynch syndrome	Møller et al. (2016)

Table 11.1 Translational Bioinformatics Resources for Cancer Studies—cont'd

Tools	Web URL	Contents	References
SomamiR 2.0	http://compbio.uthsc.edu/SomamiR	Cancer somatic mutations	Bhattacharya and Cui (2016)
SurvExpress	http://bioinformatica.mty.itesm.mx/SurvExpress	Cancer biomarker validation	Aguirre-Gamboa et al. (2013)
The Cancer Genome Atlas (TCGA)	https://cancergenome.nih.gov/	Cancer genomics	Cancer Genome Atlas Research Network et al. (2013)
The Candidate Cancer Gene Database	http://ccgd-starrlab.oit.umn.edu	Cancer driver genes from forward genetic screens	Abbott et al. (2015)
UMD TP53 mutation database	http://p53.fr	TP53 mutations in human cancer	Leroy et al. (2014)

is a knowledge base about drug targets associated with cancer, as well as cancer-associated proteins and mutations (Gohlke et al., 2016; also see Table 11.1). Mutations and Drugs Portal (MDP) is a database containing drug response data and genomic information, as well as drug combinatorial strategies (Taccioli et al., 2015).

The Cancer Proteomics Database is an integration of proteomics data and information about cell death, prostate cancer, and anticancer drugs (Arntzen et al., 2015; also see Table 11.1). The Candidate Cancer Gene Database provides a platform about cancer driver genes (Abbott et al., 2015). CancerProView is a graphical image database about cancer-associated genes and proteins (Mitsuyama and Shimizu, 2012). The platform of miREC is a database of miRNAs associated with endometrial cancer (Ulfenborg et al., 2015). SomamiR 2.0 is a database about cancer somatic mutations that may affect the microRNA–competing endogenous RNA (ceRNA) interactions (Bhattacharya and Cui, 2016).

Pancreatic Cancer Database is a comprehensive database about pancreatic cancer, especially the alterations at the mRNA, protein, and miRNA levels (Thomas et al., 2014; also see Table 11.1). BcCluster is a bladder cancer (BC) database for the research of molecular signatures associated with BC invasion (Bhat et al., 2016). The Prospective Lynch Syndrome Database provides a platform for the calculation of cumulative risks by gender, genetic variants, and age for subsequent cancer for those with Lynch syndrome in previous cancer (Møller et al., 2016).

In addition, Mouse Tumor Biology (MTB) is a database about mouse models used in the studies of human cancers (Bult et al., 2015; also see Table 11.1). The UMD TP53 mutation database focuses on TP53 mutations in human cancers (Leroy et al., 2014).

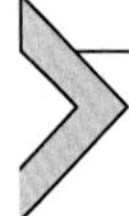

11.3 TRANSLATIONAL BIOINFORMATICS METHODS FOR CANCER STUDIES

Table 11.2 summarizes some recently developed bioinformatics methods for cancer studies. A prominent feature of the recent methodologies is the application of integrative approaches including data integration and data mining (see Chapters 3 and 4). For example, an integrated bioinformatics approach was used to study triple-negative breast cancer (TNBC). The study focused on the roles of kinase dependency in TNBC (Ryall et al., 2015). With the integration of public gene expression data, HTP pharmacological profiling data, and quantitative kinase binding data, the kinase dependency was analyzed in 12 TNBC cell lines.

Specifically, a bioinformatics method called "Kinase Addiction Ranker" was applied to query the K-Map for compounds targeting the relevant kinases (Ryall et al., 2015). The predictions were also validated using published and experimental data. Such translational approaches have revealed candidate kinases as the potential diagnostic and therapeutic targets in TNBC.

Integrative bioinformatics methods were also used for the study of the associations between breast cancer and endometriosis (Roy et al., 2015; also see Table 11.2). The methods combined environmental, epidemiological, genomic, and bioinformatics strategies for the assessments of the effects of environmental chemicals on estrogenic activities. The epidemiological correlations were examined about the influences of endocrine disrupting chemical (EDC) on health, as well as the gene–EDC interactions and disease correlations. The study revealed that several hundred genes were changed with the exposure to polychlorinated biphenyls, phthalate, or bisphenol A.

Pathway analysis indicated that the EDCs-altered genes in breast neoplasms and endometriosis were associated with the steroid hormone and inflammation pathways, especially the mitogen-activated protein kinase (MAPK) signaling pathways (Roy et al., 2015). These genes were sensitive to the environment and estrogen and could be changed in the human breast and uterine tumors, as well as the endometriosis lesions. Such identifications

Table 11.2 Examples of Translational Bioinformatics Methods for Cancer Studies

Associated Conditions	Translational Bioinformatics Methods	References
Acute myeloid leukemia	• Analysis of prognostic mutations from TCGA • Unsupervised neural network analysis for clusters and patterns	Welsh et al. (2015)
Beta-catenin implications in cancer	• Data integration from disparate sources • Literature mining • Bioinformatics Knowledge Map for protein–protein interactions, disease-associated mutations	Çelen et al. (2015)
Breast cancer	• Data-mining portal for breast cancer tissues • Genomics, methylomics, transcriptomics, proteomics, and microRNA data • Pathways in breast cancer • Links to NCBI, Ensembl, and Reactome	Cutts et al. (2015)
Breast cancer and endometriosis	• Integrated analyses for environmental and molecular links • Gene-EDC interactions and disease associations • Steroid hormone signaling and inflammation pathways	Roy et al. (2015)
Breast cancer progression	• microRNA target prediction based on differentially expressed protein-coding genes	Pinatel et al. (2014)
Cancer	• Analysis of the serine and glycine pathways using public data sets	Antonov et al. (2014)
Cancer metabolomics	• Metabolomics technologies and data generation • Data preprocessing • Multivariate data analyses, e.g., PCA, clustering, self-organizing maps	Blekherman et al. (2011)
Cutaneous metastases of prostate	• Analysis of multiple healthcare delivery networks • Aggregated EHRs	Brown et al. (2014)
Immunotherapies for cancer	• Streamline for target discovery in a bioinformatics analysis pipeline • Cataloging of potentially antigenic proteins, HLA binders, epitopes, and cotargets	Olsen et al. (2014)

Continued

Table 11.2 Examples of Translational Bioinformatics Methods for Cancer Studies—cont'd

Associated Conditions	Translational Bioinformatics Methods	References
TNBC	• Integrated analysis for kinase dependency • Integration of public gene expression data, HTP pharmacological profiling data • Kinase Addiction Ranker	Ryall et al. (2015)

EDC, endocrine disrupting chemical; *EHRs*, electronic health records; *HTP*, high-throughput; *PCA*, principal component analysis; *TCGA*, The Cancer Genome Atlas; *TNBC*, triple-negative breast cancer.

of the common environmental, molecular, and cellular risk factors in breast cancer and endometriosis at various levels would be very helpful for further discovery of systems-based biomarkers.

In a study of acute myeloid leukemia, bioinformatics was found useful for identifying prognostic mutations (Welsh et al., 2015; also see Table 11.2). Data from the TCGA database were analyzed about the cytogenetics, genetic mutations, and survival duration for prognosis. Data mining methods including unsupervised neural network and clustering analyses were performed to find the mutation or survival patterns.

The evaluations of the mutations and the clustering analysis identified several prognostic subgroups (Welsh et al., 2015). These included "good" groups that were associated with the mutations in NPM1 or TET2, "intermediate" relevant to the mutations in NPM1/DNMT3A, and "poor" relevant to the mutations in RUNX1 or FLT3-ITD/CEBPA. Such findings are meaningful for the stratification of patient subgroups for personalized medicine.

A data-mining portal of the Breast Cancer Campaign Tissue Bank (BCCTB) was constructed for the investigations of breast cancer tissues (Cutts et al., 2015; also see Table 11.2). The portal provided data integration and mining approaches for the analyses based on genomics, methylomics, transcriptomics, proteomics, and microRNA studies. The portal incorporated various resources about annotations and databases including NCBI, Ensembl, and Reactome (see Chapter 3). Such bioinformatics efforts may help save time and expenses from redundant experiments to improve the effectiveness and efficiencies of the translational processes.

Bioinformatics and computational strategies may also be used in the prediction of the correlations of microRNAs in breast cancer progression

(Pinatel et al., 2014; also see Table 11.2). For instance, the microRNA target prediction algorithms were applied to analyze the differentially expressed protein-coding genes. The mining of the cancer gene expression data sets revealed miR-223 as a potential target in breast malignancy for treatment.

In the study of cutaneous metastases of prostate cancer, bioinformatics strategies were applied to integrate information from multiple healthcare delivery networks including those from electronic health records (EHRs) (Brown et al., 2014; also see Table 11.2). The study revealed that a low rate of prostate carcinomas may lead to cutaneous metastases. The study highlighted the importance of the examinations of the complete clinical history for more accurate diagnosis of cutaneous metastases of the prostate.

In a bioinformatics study using public cancer data sets, the serine and glycine pathways in cancer cells were found significant (Antonov et al., 2014; also see Table 11.2). The expression patterns of PHGDH and SHMT2 were suggested as the prognostic factor for breast cancer associated with the predictions of patient survival outcomes. In the examinations of the patient data sets of lung cancer, the study showed that other enzymes of the pathways could also be important for prognosis. Such translational efforts may contribute to the discovery of biomarkers for human cancers.

Beta-catenin is a cell adhesion molecule and transcriptional regulator. In a recent analysis, the Bioinformatics Knowledge Map was applied to study the roles of beta-catenin in cancer (Çelen et al., 2015; also see Table 11.2). Methods of data integration were performed to understand the data from different sources to construct a bioinformatics platform to support literature mining and data mining based on biomedical ontologies and curated databases. The knowledge "maps" contained information about posttranslational modifications (PTMs), protein–protein interactions, disease-related mutations, and transcription factors associated with beta-catenin and their targets.

The study emphasized the roles of the proteins in various relation types and revealed the proteins in feedback loops associated with beta-catenin transcriptional processes (Çelen et al., 2015). The examination of the multiple networks related to PTM proteoform-specific functions identified the significance of the cyclin-dependent kinase CDK5. The analyses of the cancer-related mutation data helped in identifying the relevant patterns in different tissue types correlated with beta-catenin mutations and cancer. The integrative strategies such as those in data integration and data mining demonstrated their importance in translational bioinformatics (see Chapters 3 and 4).

Furthermore, using a bioinformatics analysis pipeline, specialized bioinformatics tools and databases were found useful for the immunotherapy target discovery for cancer (Olsen et al., 2014; also see Table 11.2). These approaches included the classifications of possible antigenic proteins, the finding of potential HLA binders, and the identifications of epitopes and co-targets. The study indicated that translational bioinformatics strategies such as those about tumor antigen HER2 may help understand drug resistance to improve the efficiencies in immunotherapy.

Moreover, bioinformatics approaches such as data preprocessing and multivariate data analysis can also be used for cancer metabolomics (Blekherman et al., 2011; also see Table 11.2). Various techniques can be helpful for such purposes, such as principal component analysis (PCA), clustering, self-organizing maps (SOMs), and discriminant function analysis. These methods can be applied to track the metabolic alterations in the cellular transformation from normal to malignant.

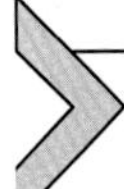

11.4 IDENTIFYING POTENTIAL SYSTEMS-BASED BIOMARKERS FOR CANCERS

11.4.1 Cancer Metastasis and Biomarkers

Metastasis is the last stage and the major factor of mortality in most cancers. The molecular mechanisms underlying metastasis are not clear, even though metastases account for most of the cancer fatalities. Available analyses of gene expression patterns in metastasis have been focusing on only a few separate genes as the "signature" biomarkers. It is critical to identify metastatic biomarkers for better understanding of this final phase of cancer progression to improve the prognosis and treatment of cancers.

To improve the understanding of the metastasis processes, systems biology approaches using HTP technologies such as microarray can be useful for analyzing gene expression data. Such analysis can help elucidate the genetic interaction networks rather than isolated genes. Bioinformatics methods such as hierarchical cluster analysis, gene ontology functional analysis, and pathway analysis can be applied for such purposes.

For example, by comparing gene expression data from various tissues including the advanced gastric cancer and adjacent noncancerous gastric tissues, metastatic tumor was found to be associated with the alterations in the apoptosis and proteasome degradation pathways (Wang et al., 2010). Specifically, the higher levels of TRAF2 and IRF3 were found in the apoptosis pathway, and increased levels of NEDD4 and UBE1 were observed in proteasome degradation pathway. Such findings can be helpful for the understanding of the cancer progression processes.

In another example, the applications of normalization and pathway analysis helped reveal that all metastatic tumors might share some properties including the alterations in energy metabolism, antigen presentation, cell adhesion, cytoskeleton remodeling, and cell cycle regulation (Ptitsyn et al., 2008). These features were common even in different tissues. In addition, compared with primary solid tumors, remarkably lower oxidative phosphorylation was observed in metastases.

Various HTP technologies including gene expression arrays, proteomics analysis, and array comparative genomic hybridization (aCGH) can be applied to analyze the prognostic expression profiles. For instance, by applying a metastatic transgenic mammary tumor model, it has been found that germline polymorphisms could be important determinants of the metastatic efficiency (Goldberger and Hunter, 2009). A concordance of survival was detected between family members with cancer, suggesting the connection between genetic inheritance factors and survival. Using aCGH and proteomic analysis, chromosomal aberrations and signaling pathways associated with the metastatic capacity have been identified (Goldberger and Hunter, 2009).

Furthermore, the analysis using activity-based proteomics helped establish the carcinoma enzyme activity profiles that could have more clinical meanings than the simple expression-based proteomics (Goldberger and Hunter, 2009). Such approaches based on the networks of interacting molecules rather than single genes can help expedite the discovery of robust and effective biomarkers for better prognosis and therapeutic results. For instance, the metabolic pathways involved in metastasis can be used as novel treatment targets.

11.4.2 Dynamical Biomarkers for Cancers

As discussed previously, the initiation and progression processes of cancers are very complex, with multiple pathways and different factors involved. It is important to elucidate the alteration tendency of carcinogenesis and the dynamic patterns of protein expressions during the different stages. Robust biomarkers need to be found based on the understanding of the associations among serum, tissues, and the microenvironment of tumors

(Berghella et al., 2011). Clinical and treatment parameters or variables at various stages should be included, from the onset to the progression of the tumor.

Specifically, in solid cancers, primary prevention and early detection may be efficient for lowering the mortality. Conventional strategies for cancer screening classify those at risk into three groups including those that are normal, and those having cancer with or without symptoms (Li et al., 2011). This static grouping method may not be helpful for lowering cancer mortality. Static genetic signatures may not be sufficient for the robust prediction of cancer progressions. Systems-based and dynamical biomarkers are needed to represent the evolving processes in cancer development.

An effort in this direction is a model of dynamic clonal evolution that can be applied as the potential biomarkers for more precise prediction and examination of cancer progression (Li et al., 2011). Such dynamical modeling and robust biomarkers for cancer development would enable timely prevention and treatment, as well as individualized administrations for better clinical outcomes.

For example, a study of different stages of colorectal cancer biopsies detected 199 differentially expressed proteins in the comparison between the the tumor, nodes, and metastasis (TNM) stages I–IV and normal tissues (Peng et al., 2012). Data mining using the SOM clustering analysis showed eight distinguished expression patterns. The proteins identified using the technologies of matrix-assisted laser desorption/ionization time-of-flight (MALDI-TOF) MS were found to be functioning in energy metabolism, acetylation, and signaling pathways (Peng et al., 2012). Using the methods of survival classifier and leave-one-out cross-validation (LOOCV) analyses, the potential prognostic biomarkers were identified with survival predictions for TNM stage I–IV patients, especially for the stage III and IV patients.

Furthermore, the cancer-associated proteins were expressed dynamically as their expression levels altered constantly throughout the tumorigenesis processes (Peng et al., 2012). Molecular indications could be observed much earlier than the observable clinical or histological alterations. Such findings demonstrate the potential applications of molecular staging profiles, as the dynamical analysis of protein expression patterns can be very useful for finding prognostic biomarkers for cancers.

To better understand the disease heterogeneity and dynamics, a systems biology approach based on parsimony phylogenetic analysis has been suggested useful for disease modeling and further biomarker discovery

(Abu-Asab et al., 2011). The strategy using parsimony phylogenetics would allow for a hierarchical classification for disease modeling. Parsimonious cladograms can be generated by using phylogenetic software such as PHYLIP's MIX.

The shared genetic expressions or differences in mutations would enable more accurate discovery of biomarkers. The parsimonious assembly of the disease heterogeneous data would facilitate the development of specimen-specific "omics" profiles. In addition, the profiling of relevant signaling pathways and molecular networks can be used as the systems-based biomarkers for disease subgroups. For example, such an analysis of prostate tumors revealed a major bifurcation in two clades showing differences between primary and metastatic prostate tumors (Abu-Asab et al., 2011).

As an example in cervical intraepithelial neoplasia (CIN), the conventional grading system based on static morphology and microscopic hematoxylin–eosin features has limitations in representing the dynamic processes as the epithelium tissues change over time (Baak et al., 2006). Functional biomarkers have been suggested useful for the evaluation of the progression and regression in individual patients, including p16, Ki-67, p53, retinoblastoma protein cytokeratin (CK)14, and CK13 (Baak et al., 2006). These biomarkers need to be detected quantitatively and separately in different layers of the epithelium to indicate the feature of a particular CIN lesion for the dynamical interpretation of the abnormal tissue for prognostic applications.

11.4.3 Examples in Breast Cancer

As a subtype of breast cancers, inflammatory breast cancer (IBC) is very destructive and difficult to treat. The genetic signatures that have been found for this complex illness cannot represent the disease well (Remo et al., 2015). For more precise diagnosis and grouping of IBC, better biomarkers are still needed to indicate the different phenotypes.

In a recent analysis, a network-based method was used to examine the master regulators (MRs) associated with the IBC pathogenesis and phenotypes (Remo et al., 2015). The study assessed the gene expression data with computational modeling and investigated the relevant cellular networks by using pathway enrichment assessments for the prediction of the possible genetic targets. These approaches were combined with the analyses using IHC and microarrays for multilevel explorations of the MRs expression patterns.

The integration of these methods help found the enriched MRs such as NFAT5 and beta-catenin that were closely associated with the IBC phenotypes (Remo et al., 2015). In addition, the NFAT5-related signaling pathways have been correlated to the IBC pathogenesis. These factors and interactions may be possible biomarkers for better diagnosis and prognosis for different disease phenotypes. Such systems-based strategies may contribute to the advancement of cancer precision medicine.

Table 11.3 shows some example of the potential biomarkers for different types of cancers. A more complete and updated list can be found at the site of Biomarkers and Systems Medicine (BSM, 2016). For instance, the phosphatidylinositol 3-kinase (PI3K)/Akt/mammalian target of rapamycin (mTOR) pathways may be useful for the prediction of sensitivity to the pathway inhibitors and can be the potential biomarkers for the subpopulation drug responses (Gonzalez-Angulo and Blumenschein, 2013). The dysregulations of the PI3K/Akt/PTEN pathways may be the possible indicators for the prognosis and high-risk prediction of node-negative breast cancer recurrence (Capodanno et al., 2009).

In addition, the insulin-like growth factor receptor 1 (IGF1R)/PI3K pathways may be important as the indications for recurrent breast cancer after tamoxifen therapy (Drury et al., 2011). The estrogen receptor alpha signaling pathways may be useful for the disease detection, prognosis, and therapeutic decisions (Ohshiro and Kumar, 2010). The HTP proteomics profiling of secretomes suggested that the IGF signaling pathways and the plasminogen activating system may be potential prognostic and predictive biomarkers for invasive breast cancer (Lawlor et al., 2009).

Furthermore, the leukemia inhibitory factor receptor (LIFR) is a breast cancer metastasis suppressor. It is associated with the microRNA miR-9 and the Hippo-YAP (transcriptional coactivator yes-associated protein) pathway. As a potential prognostic biomarker, it has been associated with breast cancer metastasis (Chen et al., 2012; also see Table 11.3).

11.4.4 Examples in Lung Cancer

In non–small cell lung cancer (NSCLC), the IGF pathways containing IGF-binding proteins (IGFBPs), such as IGFBP5 and IGFBP7, may indicate tumor progression and patient outcomes (Shersher et al., 2011; also see Table 11.3). The low levels of IGFBP5 may refer to the recurrence of the disease. The high levels of IGFBP7 may indicate the positive nodal status.

Table 11.3 Examples of Potential Systems-Based Biomarkers for Cancers

Associated Conditions	**Potential Biomarkers**	**References**
Advanced gastric cancer progression	Apoptosis pathways and proteasome degradation pathways	Wang et al. (2010)
Breast cancer drug sensitivity prediction	The PI3K/Akt/mTOR pathways	Gonzalez-Angulo and Blumenschein (2013)
Breast cancer (invasive) prognostic and prediction	The IGF signaling and the plasminogen activating systems	Lawlor et al. (2009)
Breast cancer metastasis prognosis	LIFR, microRNA miR-9, and the Hippo-YAP pathways	Chen et al. (2012)
Breast cancer (node-negative) prognosis, recurrence	The PI3K/Akt/PTEN pathways	Capodanno et al. (2009)
Breast cancer recurrence after tamoxifen therapy	The IGF1R/PI3K pathways	Drury et al. (2011)
Breast cancer surveillance, prognosis and treatment	The estrogen receptor alpha signaling pathways	Ohshiro and Kumar (2010)
Cancer metastases, therapeutic targets	The metabolic pathways	Ptitsyn et al. (2008)
Cancer therapeutic targets	Multiple pathways in redox and immune regulations, e.g., the CD30/Trx1 system	Berghella et al. (2011)
CRC prognosis	GRP78, ALDOA, CA1, and PPIA associated pathways	Peng et al. (2012)
IBC diagnosis and prognosis	The NFAT5-related signaling pathways	Remo et al. (2015)
Lung cancer diagnosis and therapeutic responses	The mTOR-signaling pathway	Ekman et al. (2012)
NSCLC tumor progression, recurrence and outcomes	The IGF pathway	Shersher et al. (2011)
NSCLC tumor heterogeneity during metastasis	The EGFR pathway	Park et al. (2009)
NSCLC prognosis	The K-Ras signaling pathway	Levallet et al. (2012)
NSCLC prognosis	The EGFR pathway	Galleges Ruiz et al. (2009)

CRC, colorectal cancer; *EGFR*, epidermal growth factor receptor; *IBC*, inflammatory breast cancer; *IGF*, insulin-like growth factor; *IGF1R*, insulin-like growth factor receptor 1; *LIFR*, leukemia inhibitory factor receptor; *mTOR*, mammalian target of rapamycin; *NSCLC*, non–small cell lung cancer; *PI3K*, phosphatidylinositol 3-kinase; *Trx1*, thioredoxin 1.

Multiple members related to the epidermal growth factor receptor (EGFR) pathways may be the potential prognostic biomarkers for NSCLC, including pERK, pSTAT3, and E-cadherin (Galleges Ruiz et al., 2009). The EGFR pathways may also indicate the heterogeneity during metastasis in NSCLC. Mutations in the relevant pathways refer to the variances between primary tumors and metastatic lymph nodes (Park et al., 2009).

In addition, the mTOR signaling pathways have been suggested as the possible diagnostic and treatment biomarker for lung cancer (Ekman et al., 2012). The β-tubulin III (TUBB3) associated with the K-Ras signaling pathways may be the possible prognostic biomarker for NSCLC treated by neoadjuvant chemotherapy such as paclitaxel- or gemcitabine-based drugs (Levallet et al., 2012; also see Table 11.3).

REFERENCES

Abbott, K.L., Nyre, E.T., Abrahante, J., Ho, Y.-Y., Isaksson Vogel, R., Starr, T.K., 2015. The Candidate Cancer Gene Database: a database of cancer driver genes from forward genetic screens in mice. Nucleic Acids Res. 43, D844–D848.

Abu-Asab, M.S., Chaouchi, M., Alesci, S., Galli, S., Laassri, M., Cheema, A.K., Atouf, F., VanMeter, J., Amri, H., 2011. Biomarkers in the age of omics: time for a systems biology approach. OMICS 15, 105–112.

Aftimos, P.G., Barthelemy, P., Awada, A., 2014. Molecular biology in medical oncology: diagnosis, prognosis, and precision medicine. Discov. Med. 17, 81–91.

Aguirre-Gamboa, R., Gomez-Rueda, H., Martínez-Ledesma, E., Martínez-Torteya, A., Chacolla-Huaringa, R., Rodriguez-Barrientos, A., Tamez-Peña, J.G., Treviño, V., 2013. SurvExpress: an online biomarker validation tool and database for cancer gene expression data using survival analysis. PLoS One 8, e74250.

Antonov, A., Agostini, M., Morello, M., Minieri, M., Melino, G., Amelio, I., 2014. Bioinformatics analysis of the serine and glycine pathway in cancer cells. Oncotarget 5, 11004–11013.

Arntzen, M.Ø., Boddie, P., Frick, R., Koehler, C.J., Thiede, B., 2015. Consolidation of proteomics data in the Cancer Proteomics database. Proteomics 15, 3765–3771.

Baak, J.P.A., Kruse, A.-J., Robboy, S.J., Janssen, E.A.M., van Diermen, B., Skaland, I., 2006. Dynamic behavioural interpretation of cervical intraepithelial neoplasia with molecular biomarkers. J. Clin. Pathol. 59, 1017–1028.

Berghella, A.M., Pellegrini, P., Del Beato, T., Ciccone, F., Contasta, I., 2011. The potential role of thioredoxin 1 and CD30 systems as multiple pathway targets and biomarkers in tumor therapy. Cancer Immunol. Immunother. 60, 1373–1381.

Bhat, A., Mokou, M., Zoidakis, J., Jankowski, V., Vlahou, A., Mischak, H., 2016. BcCluster: a bladder cancer database at the molecular level. Bladder Cancer 2, 65–76.

Bhattacharya, A., Cui, Y., 2016. SomamiR 2.0: a database of cancer somatic mutations altering microRNA-ceRNA interactions. Nucleic Acids Res. 44, D1005–D1010.

Blekherman, G., Laubenbacher, R., Cortes, D.F., Mendes, P., Torti, F.M., Akman, S., Torti, S.V., Shulaev, V., 2011. Bioinformatics tools for cancer metabolomics. Metabolomics 7, 329–343.

Brown, G.T., Patel, V., Lee, C.-C.R., 2014. Cutaneous metastasis of prostate cancer: a case report and review of the literature with bioinformatics analysis of multiple healthcare delivery networks. J. Cutan. Pathol. 41, 524–528.

BSM, 2016. Biomarkers and Systems Medicine. http://pharmtao.com/health/category/systems-medicine/biomarkers-systems-medicine.

Bult, C.J., Krupke, D.M., Begley, D.A., Richardson, J.E., Neuhauser, S.B., Sundberg, J.P., Eppig, J.T., 2015. Mouse tumor biology (MTB): a database of mouse models for human cancer. Nucleic Acids Res. 43, D818–D824.

Cancer Genome Atlas Research Network, Weinstein, J.N., Collisson, E.A., Mills, G.B., Shaw, K.R.M., Ozenberger, B.A., Ellrott, K., Shmulevich, I., Sander, C., Stuart, J.M., 2013. The Cancer Genome Atlas Pan-Cancer analysis project. Nat. Genet. 45, 1113–1120.

Capodanno, A., Camerini, A., Orlandini, C., Baldini, E., Resta, M.L., Bevilacqua, G., Collecchi, P., 2009. Dysregulated PI3K/Akt/PTEN pathway is a marker of a short disease-free survival in node-negative breast carcinoma. Hum. Pathol. 40, 1408–1417.

Çelen, İ., Ross, K.E., Arighi, C.N., Wu, C.H., 2015. Bioinformatics knowledge map for analysis of beta-catenin function in cancer. PLoS One 10, e0141773.

Chen, D., Sun, Y., Wei, Y., Zhang, P., Rezaeian, A.H., Teruya-Feldstein, J., Gupta, S., Liang, H., Lin, H.-K., Hung, M.-C., et al., 2012. LIFR is a breast cancer metastasis suppressor upstream of the Hippo-YAP pathway and a prognostic marker. Nat. Med. 18, 1511–1517.

Cutts, R.J., Guerra-Assunção, J.A., Gadaleta, E., Dayem Ullah, A.Z., Chelala, C., 2015. BCCTBbp: the Breast Cancer Campaign Tissue Bank bioinformatics portal. Nucleic Acids Res. 43, D831–D836.

Deng, X., Nakamura, Y., 2016. Cancer precision medicine: from cancer screening to drug selection and personalized immunotherapy. Trends Pharmacol. Sci. 38 (1), 15–24.

Drury, S.C., Detre, S., Leary, A., Salter, J., Reis-Filho, J., Barbashina, V., Marchio, C., Lopez-Knowles, E., Ghazoui, Z., Habben, K., et al., 2011. Changes in breast cancer biomarkers in the IGF1R/PI3K pathway in recurrent breast cancer after tamoxifen treatment. Endocr. Relat. Cancer 18, 565–577.

Ekman, S., Wynes, M.W., Hirsch, F.R., 2012. The mTOR pathway in lung cancer and implications for therapy and biomarker analysis. J. Thorac. Oncol. 7, 947–953.

Galleges Ruiz, M.I., Floor, K., Steinberg, S.M., Grünberg, K., Thunnissen, F.B.J.M., Belien, J.a. M., Meijer, G.A., Peters, G.J., Smit, E.F., Rodriguez, J.A., et al., 2009. Combined assessment of EGFR pathway-related molecular markers and prognosis of NSCLC patients. Br. J. Cancer 100, 145–152.

Gao, J., Aksoy, B.A., Dogrusoz, U., Dresdner, G., Gross, B., Sumer, S.O., Sun, Y., Jacobsen, A., Sinha, R., Larsson, E., et al., 2013. Integrative analysis of complex cancer genomics and clinical profiles using the cBioPortal. Sci. Signal. 6, l1.

Gohlke, B.-O., Nickel, J., Otto, R., Dunkel, M., Preissner, R., 2016. CancerResource–updated database of cancer-relevant proteins, mutations and interacting drugs. Nucleic Acids Res. 44, D932–D937.

Goldberger, N.E., Hunter, K.W., 2009. A systems biology approach to defining metastatic biomarkers and signaling pathways. Wiley Interdiscip. Rev. Syst. Biol. Med. 1, 89–96.

Gonzalez-Angulo, A.M., Blumenschein, G.R., 2013. Defining biomarkers to predict sensitivity to PI3K/Akt/mTOR pathway inhibitors in breast cancer. Cancer Treat. Rev. 39, 313–320.

Guhathakurta, D., Sheikh, N.A., Meagher, T.C., Letarte, S., Trager, J.B., 2013. Applications of systems biology in cancer immunotherapy: from target discovery to biomarkers of clinical outcome. Expert Rev. Clin. Pharmacol. 6, 387–401.

Huang, L., Fernandes, H., Zia, H., Tavassoli, P., Rennert, H., Pisapia, D., Imielinski, M., Sboner, A., Rubin, M.A., Kluk, M., et al., 2016. The cancer precision medicine knowledge base for structured clinical-grade mutations and interpretations. J. Am. Med. Inform. Assoc, pii:ocw148.

Lawlor, K., Nazarian, A., Lacomis, L., Tempst, P., Villanueva, J., 2009. Pathway-based biomarker search by high-throughput proteomics profiling of secretomes. J. Proteome Res. 8, 1489–1503.

Leroy, B., Anderson, M., Soussi, T., 2014. TP53 mutations in human cancer: database reassessment and prospects for the next decade. Hum. Mutat. 35, 672–688.

Levallet, G., Bergot, E., Antoine, M., Creveuil, C., Santos, A.O., Beau-Faller, M., de Fraipont, F., Brambilla, E., Levallet, J., Morin, F., et al., 2012. High TUBB3 expression, an independent prognostic marker in patients with early non-small cell lung cancer treated by preoperative chemotherapy, is regulated by K-Ras signaling pathway. Mol. Cancer Ther. 11, 1203–1213.

Li, X., Blount, P.L., Vaughan, T.L., Reid, B.J., 2011. Application of biomarkers in cancer risk management: evaluation from stochastic clonal evolutionary and dynamic system optimization points of view. PLoS Comput. Biol. 7, e1001087.

Mitsuyama, S., Shimizu, N., 2012. CancerProView: a graphical image database of cancer-related genes and proteins. Genomics 100, 81–92.

Møller, P., Seppälä, T., Bernstein, I., Holinski-Feder, E., Sala, P., Evans, D.G., Lindblom, A., Macrae, F., Blanco, I., Sijmons, R., et al., 2016. Incidence of and survival after subsequent cancers in carriers of pathogenic MMR variants with previous cancer: a report from the prospective Lynch syndrome database. Gut, pii:gutjnl-2016-311403.

Ohshiro, K., Kumar, R., 2010. Evolving pathway-driven biomarkers in breast cancer. Expert Opin. Investig. Drugs 19 (Suppl. 1), S51–S56.

Olsen, L.R., Campos, B., Barnkob, M.S., Winther, O., Brusic, V., Andersen, M.H., 2014. Bioinformatics for cancer immunotherapy target discovery. Cancer Immunol. Immunother. 63, 1235–1249.

Ow, G.S., Kuznetsov, V.A., 2016. Big genomics and clinical data analytics strategies for precision cancer prognosis. Sci. Rep. 6, 36493.

Park, S., Holmes-Tisch, A.J., Cho, E.Y., Shim, Y.M., Kim, J., Kim, H.S., Lee, J., Park, Y.H., Ahn, J.S., Park, K., et al., 2009. Discordance of molecular biomarkers associated with epidermal growth factor receptor pathway between primary tumors and lymph node metastasis in non-small cell lung cancer. J. Thorac. Oncol. 4, 809–815.

Peng, Y., Li, X., Wu, M., Yang, J., Liu, M., Zhang, W., Xiang, B., Wang, X., Li, X., Li, G., et al., 2012. New prognosis biomarkers identified by dynamic proteomic analysis of colorectal cancer. Mol. Biosyst. 8, 3077–3088.

Pinatel, E.M., Orso, F., Penna, E., Cimino, D., Elia, A.R., Circosta, P., Dentelli, P., Brizzi, M.F., Provero, P., Taverna, D., 2014. miR-223 is a coordinator of breast cancer progression as revealed by bioinformatics predictions. PLoS One 9, e84859.

Ptitsyn, A.A., Weil, M.M., Thamm, D.H., 2008. Systems biology approach to identification of biomarkers for metastatic progression in cancer. BMC Bioinform. 9 (Suppl. 9), S8.

Remo, A., Simeone, I., Pancione, M., Parcesepe, P., Finetti, P., Cerulo, L., Bensmail, H., Birnbaum, D., Van Laere, S.J., Colantuoni, V., et al., 2015. Systems biology analysis reveals NFAT5 as a novel biomarker and master regulator of inflammatory breast cancer. J. Transl. Med. 13, 138.

Roy, D., Morgan, M., Yoo, C., Deoraj, A., Roy, S., Yadav, V.K., Garoub, M., Assaggaf, H., Doke, M., 2015. Integrated bioinformatics, environmental epidemiologic and genomic approaches to identify environmental and molecular links between endometriosis and breast cancer. Int. J. Mol. Sci. 16, 25285–25322.

Ryall, K.A., Kim, J., Klauck, P.J., Shin, J., Yoo, M., Ionkina, A., Pitts, T.M., Tentler, J.J., Diamond, J.R., Eckhardt, S.G., et al., 2015. An integrated bioinformatics analysis to dissect kinase dependency in triple negative breast cancer. BMC Genom. 16 (Suppl. 12), S2.

Shersher, D.D., Vercillo, M.S., Fhied, C., Basu, S., Rouhi, O., Mahon, B., Coon, J.S., Warren, W.H., Faber, L.P., Hong, E., et al., 2011. Biomarkers of the insulin-like growth factor pathway predict progression and outcome in lung cancer. Ann. Thorac. Surg. 92, 1805–1811 discussion 1811.

Song, Q., Wang, H., Bao, J., Pullikuth, A.K., Li, K.C., Miller, L.D., Zhou, X., 2015. Systems biology approach to studying proliferation-dependent prognostic subnetworks in breast cancer. Sci. Rep. 5, 12981.

Taccioli, C., Sorrentino, G., Zannini, A., Caroli, J., Beneventano, D., Anderlucci, L., Lolli, M., Bicciato, S., Del Sal, G., 2015. MDP, a database linking drug response data to genomic information, identifies dasatinib and statins as a combinatorial strategy to inhibit YAP/TAZ in cancer cells. Oncotarget 6, 38854–38865.

Thomas, J.K., Kim, M.-S., Balakrishnan, L., Nanjappa, V., Raju, R., Marimuthu, A., Radhakrishnan, A., Muthusamy, B., Khan, A.A., Sakamuri, S., et al., 2014. Pancreatic Cancer Database: an integrative resource for pancreatic cancer. Cancer Biol. Ther. 15, 963–967.

Ulfenborg, B., Jurcevic, S., Lindlöf, A., Klinga-Levan, K., Olsson, B., 2015. miREC: a database of miRNAs involved in the development of endometrial cancer. BMC Res. Notes 8, 104.

Wang, Y.-Y., Ye, Z.-Y., Zhao, Z.-S., Tao, H.-Q., Li, S.-G., 2010. Systems biology approach to identification of biomarkers for metastatic progression in gastric cancer. J. Cancer Res. Clin. Oncol. 136, 135–141.

Welsh, K.J., Nedelcu, E., Wahed, A., Bai, Y., Dasgupta, A., Nguyen, A., 2015. Bioinformatics analysis to determine prognostic mutations of 72 de novo acute myeloid leukemia cases from the cancer Genome Atlas (TCGA) with 23 most common mutations and no abnormal cytogenetics. Ann. Clin. Lab. Sci. 45, 515–521.

Yersal, O., Barutca, S., 2014. Biological subtypes of breast cancer: prognostic and therapeutic implications. World J. Clin. Oncol. 5, 412–424.

CHAPTER TWELVE

Aging and Age-Associated Diseases: Translational Bioinformatics and Systems Biology Methods

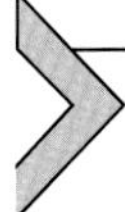

12.1 INTRODUCTION: CHALLENGES AND OPPORTUNITIES IN AGING STUDIES

Aging has complex consequences especially in age-associated diseases such as Alzheimer's disease, cancer, cardiovascular diseases, and type 2 diabetes (T2D). The worldwide increasing rates of these illnesses demand more investigations about the relationships between aging and diseases, which may also be beneficial for promoting healthy aging.

However, failed preclinical models and ineffective chemical drug candidates have made it difficult to achieve effective therapies for age-associated disorders, especially Alzheimer's disease (AD) (Flood et al., 2011). Many factors may cause such failure, including the wrong drug targets that are not closely related to the disease processes, and the lack of the translation of animal models into clinical treatments for humans.

It is urgent to identify and validate robust biomarkers to accurately represent disease onsets and progressions for better preventive strategies and drug targets. This is especially true for neurodegenerative diseases (NDs) such as AD, Parkinson's disease (PD), and amyotrophic lateral sclerosis (ALS) (Kori et al., 2016).

Translational bioinformatics may enhance the understanding in systems biology of aging with the simulation of the dynamics of biological systems in the aging processes (Mooney et al., 2016). Previous computational models have put emphasis on the detailed components but missed the full scope of aging. The conventional reductionist methods have been focusing on the separate parameters in aging. However, the complex genotypic and phenotypic alterations refer to the multiplex factors that need systemic examinations (Zierer et al., 2015).

Translational Bioinformatics and Systems Biology Methods for Personalized Medicine
ISBN 978-0-12-804328-8
http://dx.doi.org/10.1016/B978-0-12-804328-8.00012-7

The new directions focusing on the systems part would allow for the illustrations of the network interactions at various scales and levels of time and space during aging (Chauhan et al., 2015; also see Chapter 7). The systems biology-based approaches may link various aging stages at different structural, temporal, and spatial levels. The systematic analyses of the large-scale data sets such as those from the high-throughput (HTP) and "omics" data may help elucidate the complexity in the feedback loops and crosstalk among different organs and systems (see Chapter 2).

The profiling and modeling of the systemic factors can contribute to the discovery of biomarkers for the age-associated illnesses by elucidating the accumulation or "emergence" of the results from the alterations in the different sections in the biomedical system (Zierer et al., 2015; also see Chapter 2). The investigations of the nonlinear behaviors in the aging biological processes would enable beyond intuitive reasoning and more accurate models for predictions (see Chapters 1 and 2).

Specifically, translational bioinformatics and systems biology approaches may enable the integration of functional genomics data and complex molecular networks (see Chapter 4). For example, the analyses of gene regulation data and KEGG (see Chapter 3) gene sets could be applied to develop regulatory gene set networks (R-GSNs) to discover novel associations among the pathways in Alzheimer's disease (Suphavilai et al., 2015). Such GSNs would improve the understanding of the mechanisms underlying aging, illnesses, and drug perturbations.

In another example, metabolomics analysis was done based on the literature data about metabolite-disease associations among AD, PD, and ALS (Kori et al., 2016). The study identified 101 metabolites as the potential biomarkers for NDs, including the shared metabolite markers from different diseases such as creatine. The analysis of the disease-metabolite pathways emphasized the roles of membrane transporters including those of arginine and proline amino acids. The pathway enrichment analyses suggested that the metabolic pathways involving alanine, aspartate, glutamate, and purine metabolism may function to overcome insufficient glucose supply and energy crisis. These findings highlight the key roles of metabolite-based biomarkers in NDs (Kori et al., 2016).

In addition, the immune system has the essential role in aging and age-associated diseases. The terms of "immune aging" or "immunosenescence" refer to the aging routes related to the declining functions in the innate and adaptive immunity (O'Connor et al., 2014). The combination of translational bioinformatics and experimental approaches may lead to systems-based models with predictive capabilities to describe the interactions among immune molecules, cells, and tissues in aging. These would help understand the dynamical aging processes in the whole organism.

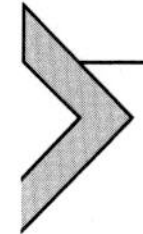

12.2 RESOURCES AND METHODS IN TRANSLATIONAL BIOINFORMATICS FOR AGING STUDIES

In addition to those discussed in Chapters 3 and 4, Table 12.1 lists some databases and bioinformatics resources that can be useful for translational studies in aging and age-related diseases. For example, Geroprotectors is a database about treatments of aging and age-related diseases (Moskalev et al., 2015; also see Table 12.1). Deep Biomarkers of Human Aging is a platform about the application of deep neural networks for the identification of biomarkers (Putin et al., 2016). GeneFriends provides an analysis tool for finding genetic targets for aging and complex diseases (van Dam et al., 2012).

AlzBase is an integrative database about genetic dysfunctions in AD (Bai et al., 2016; also see Table 12.1). The National Alzheimer's Coordinating Center (NACC) is a comprehensive platform to support exploratory and explanatory studies in AD (Beekly et al., 2007). PolyQ is a database about the studies of NDs using mouse models (Szlachcic et al., 2015).

Table 12.2 shows some examples of translational bioinformatics methods for supporting the systems biology studies of aging and associated illnesses including AD. For example, translational methods were applied for the phenotyping of Aβ sensitivity, transcriptomic profiling, and data mining of published patient data (Hadar et al., 2016; also see Table 12.2). The study found that the lower peripheral and brain expression levels of the regulator of G-protein signaling 2 (RGS2) could be a potential biomarker for early AD detection and treatment. Another study integrated bioinformatics and imaging informatics. Using methods such as whole genome sequencing (WGS) and single variant analyses, the associations were identified between the PSEN1 p. E318G variant and the risk of late-onset AD (LOAD) among APOE ε4 carriers (Nho et al., 2016).

Integrative strategies have been used for analyzing AD gene expression profiles from the resources including Gene Expression Omnibus (GEO) database, Gene Ontology (GO), and Kyoto Encyclopedia of Genes and Genomes (KEGG) pathways (Feng et al., 2015; also see Chapters 3 and 4). The study identified some significant functional roles of protein–protein interaction (PPI) networks including upregulated and downregulated genes. These networks may be considered for potential biomarkers and therapeutic targets.

Table 12.1 Translational Bioinformatics Resources for Studies of Aging and Age-Associated Diseases

Databases/Tools	Web URL	Contents	References
AlzBase	http://alz.big.ac.cn/alzBase/	Gene dysregulation in Alzheimer's disease	Bai et al. (2016)
Deep Biomarkers of Human Aging	http://www.aging.ai/	Deep neural networks for biomarker development	Putin et al. (2016)
GeneFriends	http://genefriends.org/	A coexpression analysis tool about gene targets for aging and complex diseases	van Dam et al. (2012)
Geroprotectors	http://geroprotectors.org	Therapeutic interventions in aging and age-related diseases	Moskalev et al. (2015)
National Alzheimer's Coordinating Center (NACC)	https://www.alz.washington.edu/	For Alzheimer's disease research	Beekly et al. (2007)
PolyQ database	http://conyza.man.poznan.pl/	For research of neurodegenerative diseases	Szlachcic et al. (2015)

Table 12.2 Examples of Translational Bioinformatics Methods for Studies of Age-Associated Diseases

Associated Conditions	Translational Bioinformatics Methods	References
AD	• Bioinformatics workflow with known multivariate methods • Support vector machines • Biclustering	Augustin et al. (2011)
AD	• Analysis of profiles from GEO database • Analysis of GO, KEGG pathways	Feng et al. (2015)
AD	• Genome-wide transcriptomic profiling • Data mining of published patient data	Hadar et al. (2016)
AD	• Nonlinear dynamics analyses, FD computation, entropy correlations	Holden et al. (2013)
AD	• Genome expression profiling analysis	Li et al. (2012)
AD	• Analysis of microarray data • Hierarchical clustering • Function analysis • Target genes prediction • Network construction	Zhao et al. (2016)
AD drug repurposing	• Analyses of molecular similarity • Pathway/ontology enrichment and networks	Siavelis et al. (2016)
LOAD	• WGS • Integration of bioinformatics and imaging informatics	Nho et al. (2016)
T2D, AD	• Analysis of GWAS data at SNP, gene, pathway levels • Functional enrichment analysis	Gao et al. (2016)
PD	• Analyses for disease-affected genes and pathways • Expression Data Up-Stream Analysis • Analyses of GEO data sets	Fu and Fu (2015)
PD	• Maps for common genetic variability • Analysis of the entire genome in a systematic, cost-effective way	Scholz et al. (2012)
PD	• HTP for the expression profiles	Zhang et al. (2014)

AD, Alzheimer's disease; *FD*, fractal dimension; *GEO*, Gene Expression Omnibus; *GO*, Gene Ontology; *GWAS*, genome-wide association studies; *HTP*, high-throughput; *KEGG*, Kyoto Encyclopedia of Genes and Genomes; *LOAD*, Late-onset AD; *PD*, Parkinson's disease; *SNP*, single-nucleotide polymorphism; *T2D*, type 2 diabetes; *WGS*, whole genome sequencing.

Data mining approaches including hierarchical clustering, functional analysis, target gene prediction, and interactive networks were applied for assessing AD microarray data from the expression profiles (Zhao et al., 2016; also see Chapter 4). The study found some possible target genes and relevant networks including FLT1 as the potential biomarkers. In another genome expression profiling analysis, several genes were found abnormally expressed in the metabolic and signal transduction pathways in the hippocampus (Li et al., 2012).

Bioinformatics studies about both T2D and AD examined the data from genome-wide association studies (GWAS) at various system levels including single-nucleotide polymorphisms (SNPs), genes, and pathways (Gao et al., 2016). The functional enrichment analysis identified some shared factors between T2D and AD, including the SNPs such as rs111789331 and rs66626994, as well as the significance of lipid metabolism associated pathways.

In addition, the nonlinear dynamics studies of AD including fractal dimension computation and entropy correlation analysis revealed the significance of hypocretin neuropeptide precursor (HCRT), a hypothalamus neurotransmitter associated with the wake/sleep cycle (Holden et al., 2013). Such information may be useful for AD drug discovery. Translational bioinformatics strategies may also contribute to AD drug repurposing. For instance, by examining the molecular similarities, interactions, and networks, a list of 27 potential anti-Alzheimer agents were composed (Siavelis et al., 2016).

Bioinformatics strategies emphasizing the workflow may also be important. The study using known multivariate methods such as support vector machines and biclustering helped with the discovery of significant modules associated with transcription factor families including EGRF/ZBPF (Augustin et al., 2011). The integration of in silico promoter and multivariate analyses may help elucidate the significant regulation mechanisms of genes and pathways in the multifactorial disorders.

Together with systems biology and HTP technologies, bioinformatics approaches were applied for finding disease-associated genes and pathways by assessing tissue samples from patients with PD (Fu and Fu, 2015). The study also used Expression Data Up-Stream Analysis and evaluated genomic data sets from GEO. The study revealed RNA metabolism pathology as possible factors of PD. The functional analyses showed that the dysfunctions of the transport system could be important in the early stages of the neurodegeneration processes. On the other hand, the mitochondrial dysfunctions could occur during a later phase of the disease.

In another study of PD, bioinformatics and HTP approaches were used for examining the expression profiles (Zhang et al., 2014). The study identified 181 differentially expressed genes with a similar expression trend. The analyses showed that these genes were enriched in various biological activities such as transcriptional regulations, disease progression, and drug responses. The clusters of these genes and relevant pathways may be useful for the discovery of biomarkers for the early diagnosis of PD.

Translational bioinformatics studies of PD may provide the illustration of the detailed maps for the common genetic variability (Scholz et al., 2012). Together with systems biology, such strategies may help examine the entire genome in a rapid, systematic, and inexpensive way. Using these approaches, about 30 genetic loci associated with the pathogenesis of PD were identified with the emphasis on the essential molecular pathways. These examples have demonstrated that the approaches in neurogenomics, systems biology, and translational bioinformatics can be valuable for understanding the complex age-associated diseases to support the development of rational interventions.

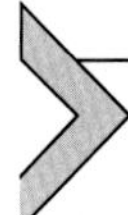

12.3 COMPREHENSIVE "OMICS" PROFILING FOR NEURODEGENERATIVE DISEASES

Age-associated NDs are influencing over 40 million people worldwide (Wood et al., 2015). As discussed earlier, the multifaceted and heterogeneous factors in the complex pathophysiology of the illnesses are hard to detect or understand. Comprehensive profiling and integrative strategies are needed to tackle the complexities.

The multidimensional investigations such as microarray and mass spectrometry technologies may provide detailed information about the onset and progression of NDs at various levels (Wood et al., 2015). The integration and mining of the data about human tissues and mouse models, together with the efforts from computational modeling, may help identify the systems-based biomarkers such as the protein aggregations in the complicated pathophysiology. Such approaches would enable the selection of new treatment targets based on the network studies of the potential biomarkers.

At the molecular level, genetic loci and polymorphisms have been associated with different neurodevelopmental and neurodegenerative illnesses (Parikshak et al., 2015). Studies using systems biology and HTP approaches would help elucidate the hierarchical context not only at the molecular level, but also the interrelationships of the cellular networks, neural circuits, and organismal cognition.

As a typical ND, AD is multifactorial associated with various "omics" including genomics, transcriptomics, metabolomics, epigenomics, interactomics, and environmental interactions (see Chapter 3). Integrative studies involving such a broad range of "omics" would enable the finding of system-based biomarkers such as altered cellular networks to represent the disease onset and development for preventive and treatment strategies during early phases.

For instance, microRNAs (miRNAs) have the essential roles in the mRNA activities in the central nervous system (Roth et al., 2016). They are critical in the gene expression profiles in the spatiotemporal dimensions. They have been closely associated with the neuronal plasticity, the aging processes, and age-associated neurodegeneration including AD.

AD has been related to genomic susceptibility, the dysfunctions of the central amyloid precursor protein (APP), and the alterations in the tau networks (Castrillo and Oliver, 2016). These alterations can cause the elevation of toxic species and the imbalances in the interactions. Among the complex factors, miRNAs and miRNA networks may be involved in the activities of APP, Aβ, and the cellular subnetworks of tau (Roth et al., 2016). The dysfunctions in these interactions have been related to the initiation and development of AD.

Other crucial factors include the homeostatic networks that can be disease counteracting, such as the quality control of proteins, proteostasis, and the protein folding chaperone networks (Castrillo and Oliver, 2016). In addition, the ubiquitin proteasome system, endolysosomal network, and various stress-response pathways are also important.

In summary, the comprehensive profiling of such interrelationships may be useful for the identification of the personalized biomarkers to represent the disease onset and progression. The systems biology modeling of the complex miRNA interactions may incorporate the findings from the transcriptomic, proteomic, metabolomic, and interactomic levels (Roth et al., 2016). The profiles of possible systems-based biomarkers may address the dysfunctions in the redox and homeostatic interactions for the early diagnosis and preventive strategies to decrease the production of toxic species. Integrative interventions may also be designed to improve the normal homeostatic reactions to manage the disease development.

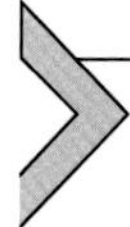

12.4 FINDING POTENTIAL SYSTEMS-BASED BIOMARKERS FOR AGING AND ASSOCIATED DISEASES

12.4.1 Examples in Aging

The aging processes include system-wide changes both functionally and structurally. To understand the mechanisms of aging, novel bioinformatics methods are needed for the identification of biomarkers of aging. Integrative and combinatorial biomarkers should be especially useful because they can be applied to quantify various processes on multiple levels of the complex biological organism.

Comprehensive biomarkers can also be helpful for the elucidation of the heterogeneities in populations. The construction of systems biology models for molecular pathways and networks would contribute to the discovery of critical and diagnostic components as candidate biomarkers for the prediction of the progression stages of aging (Kriete, 2006).

Table 12.3 lists some examples of potential systems-based biomarkers for aging and age-associated diseases. A more complete and updated list can be found at the site of Biomarkers and Systems Medicine (BSM, 2016). Some of these potential biomarkers have been discussed earlier in this chapter. In addition to those, circulating inflammatory mediators including cytokines, chemokines, growth factors, and angiogenic factors have been associated with age-related alterations. A study of 397 healthy subjects between 40 and 80 years old showed that with aging, the higher levels of serum markers were observed including interferon-γ-inducible chemokines (MIG and IP-10), eotaxin, chemoattractant for eosinophils, and soluble tumor necrosis factor receptor II (TNFR-II) (Shurin et al., 2007).

On the other hand, lower serum levels of the regulators of cell growth and differentiation were observed with aging, including EGFR and EGF (Shurin et al., 2007; also see Table 12.3). Such findings indicate that these pathways play important roles in age-associated immunosenescence and can be used as the candidate biomarkers of aging and age-related diseases.

As another example, the activators in the nuclear factor-kappa B (NF-κB) signaling pathway have been suggested as the potential biomarkers for aging and age-associated diseases, as well as the possible treatment targets (Balistreri et al., 2013; also see Table 12.3). In addition, WNT16B has been proposed as a potential biomarker associated with the p53 activity and the phosphoinositide 3-kinase (PI3K)/AKT pathways for the cellular replicative senescence (Binet et al., 2009).

Table 12.3 Examples of Potential Systems-Based Biomarkers for Aging and Age-Associated Diseases

Associated Conditions	Potential Biomarkers	References
AD diagnosis	Networks including FLT1	Zhao et al. (2016)
AD diagnosis	The profiles of toxic Aβ oligomers and tau species, the alterations in the splicing and transcriptomic patterns	Castrillo and Oliver (2016)
AD diagnosis; primary and secondary prevention/treatment	Functional imaging about the dynamical brain activation, functional connectivity of the neural networks	Prvulovic et al. (2011)
AD early detection	The heme degradation pathway (including heme oxygenase-1, biliverdin reductase A, or biliverdin reductase B)	Mueller et al. (2010)
AD dynamics	Neuronal calcium sensor proteins especially VILIP-1 in the CSF	Mroczko et al. (2015)
AD early detection and treatment	The regulators of RGS2, neuronal plasticity	Hadar et al. (2016)
AD onset and symptoms prediction	Complex dynamical biomarkers representing the temporal evolvement	Jack et al. (2013, 2010)
AD stages and severity	Cerebrospinal fluid Aβ42, p-tau, t-tau, hippocampal volumes, FDG-PET	Mouiha et al. (2012)
Aging	Interferon-γ-inducible chemokines, eotaxin, chemoattractant for eosinophils, TNFR-II	Shurin et al. (2007)
Aging and age-associated diseases therapeutic targets	NF-κB signaling pathway activators	Balistreri et al. (2013)
Cellular replicative senescence	The PI3K/AKT pathways	Binet et al. (2009)
Neurodegenerative diseases	The disease-metabolite-pathways, membrane transporters	Kori et al. (2016)
PD early diagnosis; disease progression; drug responses	The clusters of genes and relevant DEG pathways	Zhang et al. (2014)

AD, Alzheimer's disease; *Aβ*, amyloid-β; *CSF*, cerebrospinal fluid; *DEG*, differentially expressed genes; *FDG-PET*, fluorodeoxyglucose-positron emission tomography; *NF-κB*, nuclear factor-kappa B; *p-tau*, phosphorylated tau; *PD*, Parkinson's disease; *PI3K*, phosphoinositide 3-kinase; *RGS2*, regulator of G-protein signaling 2; *t-tau*, total-tau; *TNFR-II*, tumor necrosis factor receptor II; *VILIP-1*, visinin-like protein 1.

12.4.2 Dynamical Biomarkers for Alzheimer's Disease

Dynamical biomarkers representing the temporal evolvement need to be highlighted to predict the onset and development of the symptoms in AD (Jack et al., 2013; also see Table 12.3). The dynamical modeling in the identification of AD biomarkers put emphasis on the temporal factors instead of the symptom severity to illustrate the disease development and progression for personalized diagnosis and treatment.

The establishment of the accurate association between biomarkers and disease stages and severity can make it possible to predict the cognitive decline and disease development. For example, an analysis was performed on a cross-sectional data set from 576 subjects including baseline data on cerebrospinal fluid (CSF) amyloid-β (Aβ)42, phosphorylated tau (p-tau), and total tau (t-tau), hippocampal volumes, and fluorodeoxyglucose (FDG)-positron emission tomography (PET) (Mouiha et al., 2012; also see Table 12.3).

The analysis supported a local quadratic regression model and showed that the relationship between biomarkers and disease severity was nonlinear with differences among biomarkers (Mouiha et al., 2012). Such studies indicated that dynamical models of biomarkers should be established for complex diseases such as AD.

In another example, the initial conditions in AD have been related to abnormal functions of β-amyloid (Aβ) peptide with the accumulation of Aβ plaques in the brain (Jack et al., 2010). The dysfunctions may begin as early as the predisposed individuals have no abnormal clinical symptoms. At this stage, biomarkers of the brain β-amyloidosis have been suggested to have abnormal levels in CSF with higher amyloid PET tracer retention. The lagging stage after this is different from patient to patient, whereas the pathological processes are becoming dominant with neuronal dysfunction and neurodegeneration, requiring personalized prediction and diagnosis.

Synaptic dysfunction has been related to neurodegeneration, with potential indicators as lower fluorodeoxyglucose uptake on PET. Such progressive changes require a dynamical model to map AD biomarkers with various disease stages, e.g., initial abnormal Aβ biomarkers followed by neurodegenerative biomarkers and cognitive symptoms in later stages, which are associated with clinical symptom severity (Jack et al., 2010).

In a study of mild cognitive impairment in AD, the higher levels of neuronal calcium sensor proteins, especially the visinin-like protein 1 (VILIP-1) in the CSF, have been considered as a dynamic biomarker compared with those without cognitive impairment (Mroczko et al., 2015; also see Table 12.3). The higher levels of VILIP-1 were associated with lower Aβ42/40 ratio and elevated pTau181.

Another study found that the concentrations of t-tau and p-tau were much lower in late converters (5–10 years) than those in very early converters (Jack et al., 2013). It is also necessary to include the interindividual variability in cognitive impairment related to the disease progression. In addition, the specific temporal ordering of biomarkers is important for the profiling. These findings indicate that it is necessary to profile biomarkers in a dynamical and individualized context to track the pathophysiological processes to understand the mechanisms of disease progression.

Functional imaging representing the dynamical brain activation and functional connectivity of the neural networks may be used as the candidate biomarkers for diagnosis and clinical study designs (Prvulovic et al., 2011). In addition, members in the heme degradation pathway such as heme oxygenase-1, biliverdin reductase A, and biliverdin reductase B have been suggested as the potential biomarkers for the early detection of AD (Mueller et al., 2010; also see Table 12.3).

In summary, NDs including Alzheimer's disease have the chronic and nonlinear dynamic features with very complex preclinical stages. Such diseases develop over years to decades with asymptomatic stages and decompensatory processes in the brain. These silent phases have been considered as crucial for the primary and secondary prevention and treatment because the early phases before the onset of cognitive decline have the potential to be functionally reversible.

REFERENCES

Augustin, R., Lichtenthaler, S.F., Greeff, M., Hansen, J., Wurst, W., Trümbach, D., 2011. Bioinformatics identification of modules of transcription factor binding sites in Alzheimer's disease-related genes by in silico promoter analysis and microarrays. Int. J. Alzheimers Dis. 2011, 154325.

Bai, Z., Han, G., Xie, B., Wang, J., Song, F., Peng, X., Lei, H., 2016. AlzBase: an integrative database for gene dysregulation in Alzheimer's disease. Mol. Neurobiol. 53, 310–319.

Balistreri, C.R., Candore, G., Accardi, G., Colonna-Romano, G., Lio, D., 2013. NF-κB pathway activators as potential ageing biomarkers: targets for new therapeutic strategies. Immun. Ageing 10, 24.

Beekly, D.L., Ramos, E.M., Lee, W.W., Deitrich, W.D., Jacka, M.E., Wu, J., Hubbard, J.L., Koepsell, T.D., Morris, J.C., Kukull, W.A., et al., 2007. The National Alzheimer's Coordinating Center (NACC) database: the uniform data set. Alzheimer Dis. Assoc. Disord. 21, 249–258.

Binet, R., Ythier, D., Robles, A.I., Collado, M., Larrieu, D., Fonti, C., Brambilla, E., Brambilla, C., Serrano, M., Harris, C.C., et al., 2009. WNT16B is a new marker of cellular senescence that regulates p53 activity and the phosphoinositide 3-kinase/AKT pathway. Cancer Res. 69, 9183–9191.

BSM, 2016. Biomarkers and Systems Medicine. http://pharmtao.com/health/category/systems-medicine/biomarkers-systems-medicine.

Castrillo, J.I., Oliver, S.G., 2016. Alzheimer's as a systems-level disease involving the interplay of multiple cellular networks. Methods Mol. Biol. 1303, 3–48.

Chauhan, A., Liebal, U.W., Vera, J., Baltrusch, S., Junghanß, C., Tiedge, M., Fuellen, G., Wolkenhauer, O., Köhling, R., 2015. Systems biology approaches in aging research. Interdiscip. Top. Gerontol. 40, 155–176.

van Dam, S., Cordeiro, R., Craig, T., van Dam, J., Wood, S.H., de Magalhães, J.P., 2012. GeneFriends: an online co-expression analysis tool to identify novel gene targets for aging and complex diseases. BMC Genom. 13, 535.

Feng, B., Hu, P., Chen, J., Liu, Q., Li, X., Du, Y., 2015. Analysis of differentially expressed genes associated with Alzheimer's disease based on bioinformatics methods. Am. J. Alzheimers Dis. Other Demen. 30, 746–751.

Flood, D.G., Marek, G.J., Williams, M., 2011. Developing predictive CSF biomarkers-a challenge critical to success in Alzheimer's disease and neuropsychiatric translational medicine. Biochem. Pharmacol. 81, 1422–1434.

Fu, L.M., Fu, K.A., 2015. Analysis of Parkinson's disease pathophysiology using an integrated genomics-bioinformatics approach. Pathophysiology 22, 15–29.

Gao, L., Cui, Z., Shen, L., Ji, H.-F., 2016. Shared genetic etiology between type 2 diabetes and Alzheimer's disease identified by bioinformatics analysis. J. Alzheimers Dis. 50, 13–17.

Hadar, A., Milanesi, E., Squassina, A., Niola, P., Chillotti, C., Pasmanik-Chor, M., Yaron, O., Martásek, P., Rehavi, M., Weissglas-Volkov, D., et al., 2016. RGS2 expression predicts amyloid-β sensitivity, MCI and Alzheimer's disease: genome-wide transcriptomic profiling and bioinformatics data mining. Transl. Psychiatry 6, e909.

Holden, T., Nguyen, A., Lin, E., Cheung, E., Dehipawala, S., Ye, J., Tremberger, G., Lieberman, D., Cheung, T., 2013. Exploratory bioinformatics study of lncRNAs in Alzheimer's disease mRNA sequences with application to drug development. Comput. Math. Methods Med. 2013, 579136.

Jack Jr., C.R., Knopman, D.S., Jagust, W.J., Petersen, R.C., Weiner, M.W., Aisen, P.S., Shaw, L.M., Vemuri, P., Wiste, H.J., Weigand, S.D., Lesnick, T.G., Pankratz, V.S., Donohue, M.C., Trojanowski, J.Q., 2013. Tracking pathophysiological processes in Alzheimer's disease: an updated hypothetical model of dynamic biomarkers. Lancet Neurol. 12, 207–216.

Jack Jr., C.R., Knopman, D.S., Jagust, W.J., Shaw, L.M., Aisen, P.S., Weiner, M.W., Petersen, R.C., Trojanowski, J.Q., 2010. Hypothetical model of dynamic biomarkers of the Alzheimer's pathological cascade. Lancet Neurol. 9, 119–128.

Kori, M., Aydın, B., Unal, S., Arga, K.Y., Kazan, D., 2016. Metabolic biomarkers and neurodegeneration: a pathway enrichment analysis of Alzheimer's disease, Parkinson's disease, and amyotrophic lateral sclerosis. OMICS 20, 645–661.

Kriete, A., 2006. Biomarkers of aging: combinatorial or systems model? Sci. Aging Knowledge Environ. 1, pe1.

Li, Y., Wu, Z., Jin, Y., Wu, A., Cao, M., Sun, K., Jia, X., Chen, M., 2012. Analysis of hippocampal gene expression profile of Alzheimer's disease model rats using genome chip bioinformatics. Neural Regen. Res. 7, 332–340.

Mooney, K.M., Morgan, A.E., Mc Auley, M.T., 2016. Aging and computational systems biology. Wiley Interdiscip. Rev. Syst. Biol. Med. 8, 123–139.

Moskalev, A., Chernyagina, E., de Magalhães, J.P., Barardo, D., Thoppil, H., Shaposhnikov, M., Budovsky, A., Fraifeld, V.E., Garazha, A., Tsvetkov, V., et al., 2015. Geroprotectors.org: a new, structured and curated database of current therapeutic interventions in aging and age-related disease. Aging (Albany NY) 7, 616–628.

Mouiha, A., Duchesne, S., Alzheimer's Disease Neuroimaging Initiative, 2012. Toward a dynamic biomarker model in Alzheimer's disease. J. Alzheimers Dis. 30, 91–100.

Mroczko, B., Groblewska, M., Zboch, M., Muszyński, P., Zajkowska, A., Borawska, R., Szmitkowski, M., Kornhuber, J., Lewczuk, P., 2015. Evaluation of visinin-like protein 1 concentrations in the cerebrospinal fluid of patients with mild cognitive impairment as a dynamic biomarker of Alzheimer's disease. J. Alzheimers Dis. 43, 1031–1037.

Mueller, C., Zhou, W., Vanmeter, A., Heiby, M., Magaki, S., Ross, M.M., Espina, V., Schrag, M., Dickson, C., Liotta, L.A., Kirsch, W.M., 2010. The heme degradation pathway is a promising serum biomarker source for the early detection of Alzheimer's disease. J. Alzheimers Dis. 19, 1081–1091.

Nho, K., Horgusluoglu, E., Kim, S., Risacher, S.L., Kim, D., Foroud, T., Aisen, P.S., Petersen, R.C., Jack, C.R., Shaw, L.M., et al., 2016. Integration of bioinformatics and imaging informatics for identifying rare PSEN1 variants in Alzheimer's disease. BMC Med. Genom. 9 (Suppl. 1), 30.

O'Connor, J.-E., Herrera, G., Martínez-Romero, A., de Oyanguren, F.S., Díaz, L., Gomes, A., Balaguer, S., Callaghan, R.C., 2014. Systems biology and immune aging. Immunol. Lett. 162, 334–345.

Parikshak, N.N., Gandal, M.J., Geschwind, D.H., 2015. Systems biology and gene networks in neurodevelopmental and neurodegenerative disorders. Nat. Rev. Genet. 16, 441–458.

Prvulovic, D., Bokde, A.L., Faltraco, F., Hampel, H., 2011. Functional magnetic resonance imaging as a dynamic candidate biomarker for Alzheimer's disease. Prog. Neurobiol. 95, 557–569.

Putin, E., Mamoshina, P., Aliper, A., Korzinkin, M., Moskalev, A., Kolosov, A., Ostrovskiy, A., Cantor, C., Vijg, J., Zhavoronkov, A., 2016. Deep biomarkers of human aging: application of deep neural networks to biomarker development. Aging (Albany NY) 8, 1021–1033.

Roth, W., Hecker, D., Fava, E., 2016. Systems biology approaches to the study of biological networks underlying Alzheimer's disease: role of miRNAs. Methods Mol. Biol. 1303, 349–377.

Scholz, S.W., Mhyre, T., Ressom, H., Shah, S., Federoff, H.J., 2012. Genomics and bioinformatics of Parkinson's disease. Cold Spring Harb. Perspect. Med. 2, a009449.

Shurin, G.V., Yurkovetsky, Z.R., Chatta, G.S., Tourkova, I.L., Shurin, M.R., Lokshin, A.E., 2007. Dynamic alteration of soluble serum biomarkers in healthy aging. Cytokine 39, 123–129.

Siavelis, J.C., Bourdakou, M.M., Athanasiadis, E.I., Spyrou, G.M., Nikita, K.S., 2016. Bioinformatics methods in drug repurposing for Alzheimer's disease. Brief. Bioinform. 17, 322–335.

Suphavilai, C., Zhu, L., Chen, J.Y., 2015. A method for developing regulatory gene set networks to characterize complex biological systems. BMC Genom. 16 (Suppl. 11), S4.

Szlachcic, W.J., Switonski, P.M., Kurkowiak, M., Wiatr, K., Figiel, M., 2015. Mouse polyQ database: a new online resource for research using mouse models of neurodegenerative diseases. Mol. Brain 8, 69.

Wood, L.B., Winslow, A.R., Strasser, S.D., 2015. Systems biology of neurodegenerative diseases. Integr. Biol. (Camb.) 7, 758–775.

Zhang, Y., Yao, L., Liu, W., Li, W., Tian, C., Wang, Z.Y., Liu, D., 2014. Bioinformatics analysis raises candidate genes in blood for early screening of Parkinson's disease. Biomed. Environ. Sci. 27, 462–465.

Zhao, Y., Tan, W., Sheng, W., Li, X., 2016. Identification of biomarkers associated with Alzheimer's disease by bioinformatics analysis. Am. J. Alzheimers Dis. Other Demen. 31, 163–168.

Zierer, J., Menni, C., Kastenmüller, G., Spector, T.D., 2015. Integration of "omics" data in aging research: from biomarkers to systems biology. Aging Cell 14, 933–944.

INDEX

'*Note*: Page numbers followed by "f" indicate figures and "t" indicate tables.'

D

E

F

G

O

P

R

S

T

U

V

W

Printed in the United States
By Bookmasters